Undergraduate Texts in Mathematics

Readings in Mathematics

Editors

S. Axler
F. W. Gehring
K. A. Ribet

T0075897

Springer Science+Business Media, LLC

Graduate Texts in Mathematics
Readings in Mathematics

Ebbinghaus/Hermes/Hirzebruch/Koecher/Mainzer/Neukirch/Prestel/Remmert: *Numbers*
Fulton/Harris: *Representation Theory: A First Course*
Remmert: *Theory of Complex Functions*
Walter: *Ordinary Differential Equations*

Undergraduate Texts in Mathematics
Readings in Mathematics

Anglin: *Mathematics: A Concise History and Philosophy*
Anglin/Lambek: *The Heritage of Thales*
Bressoud: *Second Year Calculus*
Hairer/Wanner: *Analysis by Its History*
Hämmerlin/Hoffmann: *Numerical Mathematics*
Isaac: *The Pleasures of Probability*
Laubenbacher/Pengelley: *Mathematical Expeditions: Chronicles by the Explorers*
Samuel: *Projective Geometry*
Stillwell: *Numbers and Geometry*
Toth: *Glimpses of Algebra and Geometry*

Richard Isaac

The Pleasures of Probability

With 17 Illustrations

Springer

Richard Isaac
Department of Mathematics
Lehman College, City University of New York
Bronx, NY 10468-1589
USA

The Graduate Center
33 West 42nd Street
New York, NY 10036
USA

Editorial Board

S. Axler
Mathematics Department
San Francisco State
  University
San Francisco, CA 94132
USA

F.W. Gehring
Mathematics Department
East Hall
University of Michigan
Ann Arbor, MI 48109
USA

K.A. Ribet
Mathematics Department
University of California
  at Berkeley
Berkeley, CA 94720-3840
USA

Mathematics Subject Classification (2000): 60-01

Library of Congress Cataloging-in-Publication Data
Isaac, Richard.
    The pleasures of probability / Richard Isaac.
      p. cm. — (Undergraduate texts in mathematics. Readings in
  mathematics)
    Includes bibliographical references and index.
    ISBN 978-1-4612-6912-0   ISBN 978-1-4612-0819-8 (eBook)
    DOI 10.1007/978-1-4612-0819-8
    1. Probabilities. I. Title. II. Series.
  QA273.173   1995
  519.2—dc20   94-39140

Printed on acid-free paper.

© 1995 Springer Science+Business Media New York
Originally published by Springer-Verlag New York, Inc. in 1995
Softcover reprint of the hardcover 1st edition 1995
All rights reserved. This work may not be translated or copied in whole or in part without the written permission of the publisher, Springer Science+Business Media, LLC
except for brief excerpts in connection with reviews or scholarly analysis. Use in connection with any form of information storage and retrieval, electronic adaptation, computer software, or by similar or dissimilar methodology now known or hereafter developed is forbidden.

The use of general descriptive names, trade names, trademarks, etc., in this publication, even if the former are not especially identified, is not to be taken as a sign that such names, as understood by the Trade Marks and Merchandise Marks Act, may accordingly be used freely by anyone.

Production managed by Hal Henglein; manufacturing supervised by Jacqui Ashri.
Photocomposed copy prepared from the author's TeX file.

9 8 7 6 5 4

ISBN 978-1-4612-6912-0

To the memory of my parents

Preface

The ideas of probability are all around us. Lotteries, casino gambling, the almost non-stop polling which seems to mold public policy more and more—these are a few of the areas where principles of probability impinge in a direct way on the lives and fortunes of the general public. At a more removed level there is modern science which uses probability and its offshoots like statistics and the theory of random processes to build mathematical descriptions of the real world. In fact, twentieth-century physics, in embracing quantum mechanics, has a world view that is at its core probabilistic in nature, contrary to the deterministic one of classical physics. In addition to all this muscular evidence of the importance of probability ideas it should also be said that probability can be lots of fun. It is a subject where you can start thinking about amusing, interesting, and often difficult problems with very little mathematical background.

In this book, I wanted to introduce a reader with at least a fairly decent mathematical background in elementary algebra to this world of probability, to the way of thinking typical of probability, and the kinds of problems to which probability can be applied. I have used examples from a wide variety of fields to motivate the discussion of concepts. Each chapter contains a number of such problems and applications, many of these related to gambling and games of chance, an important and picturesque source of probability thinking. I have explained the problems and the ideas they generate in what I hope is an intuitive way understandable to beginners, using almost entirely elementary algebra. One of the glories of the subject is that many of its fundamental concepts are intuitively appealing and accessible to non-specialists, especially if you look at them in the "right" way, often in a gambling or betting framework. In particular, I have tried to explain the basic ideas and importance behind two of the most profound themes in probability that thread their way throughout this book: the circle of ideas concerning the Law of Large Numbers and The Central Limit Theorem.

I see several different possible audiences for this book. The general reader, armed with basic algebra, interest, and some perseverance, can enjoy the problems and pick up as much of the theory as desired. If such a reader comes away from the book with an improved ability to assess the chances of winning a fortune by casino gambling or playing the lottery, I will have already succeeded in my aim. At another level, students, scientists and mathematicians who want an elementary overview that still manages to tread into deep waters in an informal way will find it useful and, I hope, entertaining. The book could also be used as a text for a first course in probability or as a companion to a text. But this book is not a text in the conventional sense. There is a lot of non-rigorous, intuitive argument, no attempt to be comprehensive, and the style, in its freewheeling and informal treatment of topics I find important or beautiful, is perhaps more suitable to a book of essays than a textbook with its strict structure and aims. Nevertheless, the range of topics covered includes most of those in a standard elementary introduction. But rather than give you a text-like agenda of "things you have to learn," my goals are more playful; my rallying cry is "enter this world and see the fascinating things it has to offer."

To get the most from the book the reader should be able to do basic elementary algebra fairly competently. In addition, there should be a willingness to think hard about what is being said and to check out assertions with paper and pencil. If the reader wishes to read more passively by eschewing paper and pencil and perhaps hard thought, that is all right too, for a general sense of what is going on. Perhaps a second reading will be more aggressive. Although elementary algebra is used almost entirely in arguments (there are a few places where calculus is mentioned briefly and the ideas explained), readers who bring more background in mathematics and science to the book are likely to get more out of it. For example, those who are familiar with a programming language can use the algorithms described in Chapter 14 to test probability ideas using the computer.

I am, of course, indebted to many books and articles on probability. Many of the examples discussed in this book are classical, and can be read about in a number of places, although I sometimes spend more time analyzing these problems than other sources do. Each chapter ends with a few problems for the reader to try; some are very easy, others a little harder. The answers to all problems are given at the end of the book.

Morton D. Davis and Paul R. Meyer read the manuscript and made many valuable corrections and suggestions. They were very helpful in getting me to clarify obscure presentations and think more carefully about what I was saying in many places. Of course, any errors or obfuscations in this finished product are entirely my responsibility. Shaul Foguel also offered useful comments. Richard Mosak and Melvin Fitting helped me use the LaTeX document preparation system. Melvin Fitting offered a wealth of information about preparing illustrations. Esther Phillips aided in creating some of these illustrations. To all of these people I owe much thanks. I also want to thank the people at Springer-Verlag and especially my editor, Ina Lindemann, for their help. Finally, I thank my wife, Anna, for her encouragement and support throughout this project.

Richard Isaac

Contents

1
Cars, Goats, and Sample Spaces

> Behold, there stand the caskets noble prince:
> If you choose that wherein I am contain'd,
> Straight shall our nuptial rites be solemnized:
> But if you fail, without more speech, my lord,
> You must be gone from hence immediately.
>> William Shakespeare, *Portia* in *The Merchant of Venice*

1.1 Getting your goat

It's a critical moment for you. The master of ceremonies confronts you with three closed doors, one of which hides the car of your dreams, new and shiny and desirable. Behind each of the other two doors, however, is standing a pleasant but not so shiny and somewhat smelly goat. You will choose a door and win whatever is behind it. You decide on a door and announce your choice, whereupon the host opens one of the other two doors and reveals a goat. He then asks you if you would like to switch your choice to the unopened door that you did not at first choose. Is it to your advantage to switch (assuming, of course, that you are after the car, not the goat)?

This popular puzzler created a stir in 1991 when it appeared in the newspaper and (see [32] [1]) received a lot of wrong answers from readers, even from some who were mathematicians. How do we think about a problem

---

[1]Numbers in square brackets refer to the references at the end of the book.

like this, and why is it so tricky? (The most common wrong answer was that switching is irrelevant because each of the two unopened doors would hide the car with equal probability.) What I'd like to do is use this problem to introduce you to the branch of mathematics called probability. When you finish this chapter you should be able to think about the car-goat problem and many other probability problems in a reasonable way. Let's begin with a capsule description of probability theory, its significance and history, in a few action-packed paragraphs.

1.2   Nutshell history and philosophy lesson

Probability can be described as the mathematical theory of uncertainty. Its origins are undoubtedly ancient, since an early cave dweller looking at the sky for some clue about the weather was using primitive notions of probability. In fact, one could argue that all of us use probability daily to assess risks; the probabilities used are rough estimates based on previous experience. (Dark clouds today mean rain is likely since it has rained in the past when the clouds had that look. Better carry an umbrella.) Primitive or instinctive probability, however, is very different from a developed mathematical discipline. Officially, probability as a formal theory is sometimes said to have begun in the seventeenth century with a famous correspondence between the two French mathematicians Blaise Pascal and Pierre Fermat. The gambling halls of Paris were giving life to the new science. In a sense, a casino is almost a perfect laboratory of probability in action; a serious gambler has to have a pretty good idea of the risks in order to bet rationally. After a while the gambler either has to become a mathematician or consult one.

From these somewhat frivolous beginnings, the theory developed to its present status, with applications to all branches of science, technology, and even to that citadel of uncertainty, the stock market. Moreover, the twentieth century provided a new and rather startling star role for probability ideas within the framework of modern physics. In the physics of the eighteenth century, Newton's era, it was supposed that if you only had all the data you could use the equations of physics to predict the position and velocity of a particle exactly. Physicists viewed probability as a useful tool, mainly because it was often too hard to get all the data input for a problem. So probability was tolerated, in a sense, as a lesser discipline, because if our ignorance were only eliminated we wouldn't need probability, there wouldn't be any uncertainty, it was argued. For example, if we knew all about how a coin was tossed, the accelerations, angles, and forces involved, we could in principle predict whether a head or tail would come up. That was fine, until the new physics came along and Werner Heisenberg postulated the "uncertainty principle," that for very small particles it was impossible to know both the position and velocity exactly; the better you know the position, the fuzzier your idea of the velocity becomes, and vice versa, and there isn't anything you can do about it. This idea revolutionized the foundations of physics. Here was Heisenberg now saying that in principle you could not make exact predictions; the best you could do would be to make probability statements no matter how much data you collected. It was all very distressing to Einstein, who rejected Heisenberg's theory with his famous statement "God does not play dice." However, modern physicists now believe that Heisenberg was correct.

1.3   Let those dice roll. Sample spaces

Let's begin rolling dice, tossing coins, and other such things, because this is where the heart of probability lies. Since probability measures uncertainty, we have to measure something, and these objects probabilists like to call *events* since this is a reasonable name to give to the something that happens. Now suppose we are interested in what happens when we roll a pair of dice once. Assume one die is red and the other is green. When the red die falls it can come up in six ways, and the same holds for the green die. Each possible result can be represented by an ordered pair (a, b), where a is one of the numbers from 1 to 6 and represents what the red die's number is, and b is also one of the numbers 1 to 6 representing the green die's number. So what actually happens when you roll the pair of dice once? Well, there are 36 ordered pairs (a, b) where a and b vary between 1 and 6 (just write these all out to see that for $a = 1$ there are six possibilities for b, for $a = 2$, another six possibilities for b, etc.). What happens when you roll the dice can be conveniently described by exactly one of the 36 ordered pairs possible. Each of those 36 possible ordered pairs we call an *outcome*. Outcomes are the simplest kind of event. More complicated events contain a number of outcomes. For example, the event defined by the phrase "rolling a seven" contains six outcomes; it can be described as the event

$$A = \{(1, 6), (2, 5), (3, 4), (4, 3), (5, 2), (6, 1)\}.$$

Here the curly braces tell us to regard the enclosed items as being lumped together to form the event called A. We can say that the "experiment" of rolling a pair of dice once gives rise to a *sample space* S, which is just the set of all 36 of the possible outcomes, and that any event is simply a set of some collection of these 36 elementary outcomes or building blocks for events. For instance, the event

$$B = \{(1, 1), (1, 2), (1, 3)\}$$

could be described in words as "rolling 1 with the red die and rolling between 1 and 3 inclusive with the green die." Figure 1.1 shows how the sample space S and the event B can be represented in a sketch, with the

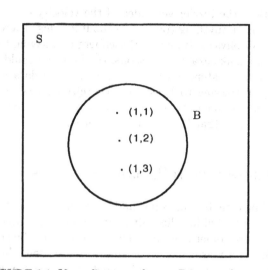

FIGURE 1.1. Venn diagram of a set B in sample space S

outcomes designated as points in the picture. Such a pictorial representation of sets is called a *Venn diagram.*

As we have seen above, an event is just a suggestive word probabilists use to talk about a set, namely, a collection of objects which, in the probability situation, is a collection of outcomes from a random experiment (the word *random* here means you can't predict the outcome in advance). As another example, the experiment of tossing a coin twice gives us a sample space S with four outcomes where, if we use H and T for head and tail, respectively, we can write:
$$S = \{(H,H),(H,T),(T,H),(T,T)\}$$
where the first entry in the ordered pair represents what happens on the first toss and the second entry what happens on the second toss. The event C, "at least one head occurs," can be written, for example, as

$$C = \{(H,H),(H,T),(T,H)\}.$$

There are a number of important points we should keep in mind about sample spaces. First, we used the word "experiment" to describe the happening that gives rise to the sample space. Experiments are usually things that can be repeated, and this is appropriate. That is because we will be considering probabilities for the most part for the kind of events that can arise only from some repeatable circumstance such as rolling dice or tossing coins. Suppose we are at a trial by jury; we will not consider an event like "the defendant is guilty" to be the kind of event to which we are going to attach a probability (at least for the moment) because it is not the kind of event arising from a repeatable experiment like rolling dice or tossing coins.

Another point is that a sample space provides what is called a *mathematical model* of the real-life situation for which it is supposed to be an abstraction. The reason for constructing this abstraction is that mathematical analysis can only be performed on the ideal structure of the sample space, not on the real-life situation itself. Once you have this model you may derive some nice mathematical relationships about the ideal structure, the abstraction. Since the abstraction resembles the real world, you might think that the mathematical relationships you found say something about the real world. You can now perform scientific experiments to check out the real-world situation. If you were clever and lucky, the mathematical model helped you decipher the real world; you know this because the results of your experiments are consistent with the mathematical relationships you obtained from the model. It could also happen that your model was too simple or otherwise in error and did not give a true picture of the real-world situation. In this case, the mathematical relationships, while true for the model, cannot be verified by laboratory experiment. Then it's back to the drawing board to look for a more accurate model.

It follows that since a sample space is constructed to model a real-life situation and is therefore only a construct, a figment of the imagination of the observer of that situation, it depends on what that observer thinks is important. For example, let us say that every now and then when you roll the dice, your dog jumps up on the table and grabs the red die in his jaws and runs under the couch with it. If you wanted, you could consider the sample space including with the 36 outcomes in S another six outcomes which could be represented as $(D, 1), (D, 2), (D, 3), (D, 4), (D, 5), (D, 6)$. Here $(D, 5)$, for example, means that the dog has run off with the red die so no number has turned up on it but the green die came up with 5. Similarly, if the dog occasionally runs off with the green die or with both dice and we want to include sample points for these occurrences, we could add points to denote this (the geometric word "point" is a convenient and suggestive word for an outcome; it derives from the practice of drawing a picture of a sample space as in Fig 1.1, with the list of all possible outcomes as a scattering of dots or points inside it). The sample space representing what happens when a pair of dice is rolled is therefore not unique; it may be considerably more complicated than the one originally given by S. It all depends on what the problem is and what you judge to be the relevant information.

1.4   Discrete sample spaces. Probability distributions and spaces

So far, as you have noticed, we don't have the idea of probability at all in our mathematical structure, the sample space. All we have is the list of all possible outcomes that can be generated by the performance of some repeatable actual or mind experiment. Then we consider sets of these outcomes to form the objects whose uncertainty we are interested in measuring. I should also say that for the moment let us restrict ourselves to what are called *discrete* sample spaces. These are sample spaces where the outcomes can be counted off using the positive integers. This includes all sample spaces with a finite number of outcomes like the ones considered above as well as certain infinite sample spaces.

Here is an example of an infinite discrete sample space which will be of great interest to us later: imagine that you have a coin and you are going to toss it repeatedly in successive trials until you get a head for the first time, and then you are going to stop. For the purposes of the problem, you are immortal as is the universe; if after a million years the coin has still not come down a head you and the universe will still be there to experience another toss. The sample space of this experiment can be represented by

$$S = \{(H), (T,H), (T,T,H), (T,T,T,H), \cdots\}.$$

Here the first term, (H), represents the outcome of getting a head on the first toss and then stopping, the second term, (T, H), the outcome of getting a tail on the first toss, a head on the second, and then stopping, etc., where the dots express that this sequence goes on ad infinitum. If n is any positive integer, S has an nth term given by an n-tuple of $n - 1$ tails followed by a terminal head; this represents the outcome that can be described by "a head for the first time at trial n." Since there is no largest value for n, there is an outcome for each positive integer. S is a discrete sample space because the elements of S are in one-to-one correspondence with the positive integers; namely, you can count them all off and not have any left over when you are done.

What is an example of a sample space that is not discrete? Consider the set W of all non-terminating decimal expansions of the form $.a_1a_2a_3 \cdots$ where the entries are any of the ten digits 0 through 9. We can think of W geometrically as representing the numbers in the interval from 0 to 1. Any such number can be written uniquely by giving its decimal expansion (the representation is not quite unique since terminating decimals like $.5 = .5000 \cdots$ have another representation as $.4999 \cdots$. The representation becomes unique if we always agree to choose the expansion ending, say, in zeros). W can be considered as a sample space by thinking of the experiment of choosing a point from the interval 0 to 1. It can be shown that W is not a discrete sample space. There are just too many points in an interval to be able to count them all off as first, second, etc., using the positive integers. The sample space W turns out to be useful for many problems. We will return to it later.

And now, finally, probability is about to make her entrance. (Tyche, the Greek goddess of chance, was, of course, a female.) Start with any discrete sample space, for example, S, the list of the 36 outcomes of rolling a pair of dice. To each of the outcomes in such a sample space associate a number between 0 and 1 such that the sum of all these numbers over all outcomes is equal to 1. The number associated with a particular outcome is called the probability of that outcome, and the entire assignment is called a *probability distribution, probability measure* or *probability mass* on S. Now we can define the probability of any event A. If A is the event with no outcomes in it (namely, what is called the *empty* set), let its probability be 0; otherwise, let its probability be the sum of the probabilities of all outcomes in the event. So, given the probability distribution on S, we can figure out the probabilities of all events in S.

The preceding paragraph tells you how to set up a probability distribution on a discrete sample space; there are an infinite number of ways to do this—as many ways as there are of assigning numbers between 0 and 1 to the outcomes such that the sum of the numbers over all outcomes is 1. But this doesn't answer the question about how to find a *useful* probability distribution in a particular problem. The usefulness of a probability distribution is not a mathematical question; it is determined by what you want the sample space to model in the real world. The particular application suggests the appropriate distribution.

Here is an important example using the sample space S of the 36 pairs of numbers (a, b), where a and b are both numbers between 1 and 6 inclusive. If S is modelling the rolls of a pair of dice, it usually seems natural to assign the number 1/36 to each of the 36 outcomes of S. This gives us what is called the *uniform* or *equally likely* distribution. It corresponds to a real-life situation in which we feel that no outcome is favored over any other outcome. This is often a reasonable way to feel about dice; they have been constructed (we believe) with physical properties and symmetries that present quite a strong case for having us assign to each outcome the same degree of uncertainty. Since the uncertainties must add to 1, each outcome is assigned uncertainty 1/36. In the case of n outcomes, the uniform distribution assigns probability $1/n$ to each outcome. Now let us calculate the probability of "rolling a seven," namely, $P(A)$, the probability of the event A as defined in Section 1.3 (we use $P(X)$ to denote the probability of the general event X). The event A contains six outcomes, and the sum of the probabilities of the six gives 6/36=1/6. In similar fashion the set B defined in Section 1.3 has probability 3/36=1/12, and the probability of any event in S can be calculated using the rule of adding up the probabilities of the outcomes that make up the event.

Let us summarize: a sample space is a mathematical model of a real or mind experiment which we imagine can be repeated under similar conditions (like rolling dice or tossing coins). If the sample space is discrete, a probability distribution can be defined on the sample space by associating with each one of the discrete outcomes a number between 0 and 1 which we call the probability of the outcome. The sum of the probabilities of all of the outcomes must add up to 1. The probability of a more general event is defined by adding up the probabilities of all the outcomes comprising

the event. An event is just a set in the sample space—the words *event* and
set are used interchangeably. The sample space with its probability distribution is sometimes called a *probability space*. The event with no elements,
the empty event, has probability 0, and the sample space, sometimes called
the *sure* event, has probability 1. The probability of an event is therefore a
weighting of the event by means of the probability distribution—a "light"
event is one of low probability; a "heavy " event is one of high probability. The heavier an event is, namely, the higher the probability, the less
uncertain is the event.

As an example, let x be one of the numbers between 2 ("snake eyes"
in gambling parlance) and 12, and suppose you and I are playing a game
whereby I will give you a dollar if x comes up on one roll of a pair of dice
(that is, x is the sum of the faces) and you give me a dollar otherwise. If
you are able to choose the value of x, it is to your advantage to choose
$x = 7$ since $P(x = 7) = 1/6$ and the probability of any other value coming
up is less than $1/6$, as you can calculate quite easily. Your choice of $x = 7$
in this game, that is, your belief that events of higher probability are less
uncertain than events of lower probability and therefore better to bet on,
is not an immediate or obvious consequence from the abstract discussion
of probability spaces as just described. That discussion only tells us how
to *calculate* probabilities, not how to interpret what they mean. But we
will see when we come to the Law of Large Numbers that the probability
of an event has a remarkable relationship with the relative frequency of
its occurrence—seven will come up roughly $1/6$ of the total number of
rolls if you perform a large number of repeated rolls, and snake eyes will
come up only about $1/36$ of the time, if the uniform distribution model
is a good one. The Law of Large Numbers will justify our intuition about
probabilities and relative frequencies. But until then we will rely on our
intuition in supposing events of higher probability to be better bets than
ones of lower probability.

1.5   The car-goat problem solved

We are ready to solve the car and goat problem. What we have to do
is construct a sample space to model the experiment. But first we must
know exactly what the experiment is in this case. That means we have to
translate the somewhat fuzzy and ambiguous phrasing of the problem into
an exact mathematical description. Let us make the problem more precise
by assuming you do indeed decide to switch, whatever happens, and let's
see what this leads to. The situation can be abstracted as follows. The game
consists of three actions: (a) first you make your initial choice of one of the
three possible doors, (b) the host chooses one of the other two doors with
a goat behind it, (c) you switch your choice. Now suppose the door with
the car behind it is labelled 1, and the remaining two doors with goats are

labelled 2 and 3. Let us describe a typical outcome from this game by a 4-tuple (u, v, w, x), where u will be the label of the door you initially choose, v is the label of the door the host opens, w is the label of the door you switch to, and x stands for the letter "W" or "L" depending on whether you win or lose the car. For example, the 4-tuple $(1, 2, 3, L)$ is shorthand for "you choose door 1 (with the car behind it), the host opens door 2, and since you switch you must switch to 3, thereby losing the car." The sample space S can be written

$$S = \{(1, 2, 3, L), (1, 3, 2, L), (2, 3, 1, W), (3, 2, 1, W)\}.$$

If you choose doors 2 or 3 initially, the rules of the game must lead to your winning; this is seen in the third and fourth outcomes of S. If you choose door 1 initially you must lose, although there are two different ways to lose depending upon which door the host opens; this is seen from the first and second outcomes of S. (By the way, the use of the L or W in the fourth place in the 4-tuples above is just a convenience. It helps us see at a glance which of the outcomes lead to a loss and which lead to a win. An entirely equivalent sample space would consist of the four triples formed by taking each outcome in S and chopping off the L or W in the last component.) Please do not read any further until you are convinced that the above four outcomes in S are the only possible ways the "game" we are describing can be played.

So far, so good. We have a probability space, but how do we get a reasonable probability distribution for it? Well, we are making a mathematical model of a real-life situation, so we have to go back to the real-life situation and ask ourselves what kind of assumptions might be realistic here. If you are in front of the three doors before you make your initial choice, on what are you going to base your decision on which door to choose? Assume you have no reason to favor any one door over any other door (you do not hear the shuffling of animal haunches nor do you smell any suspicious goatlike odors behind any particular door). This means you will probably guess at random. In probability problems, "at random" is a code phrase meaning you should choose all outcomes equally likely; that is, you assume a uniform distribution. In this case, if you had a three-sided die with the numbers $1, 2, 3$ on the faces, you would roll the die and choose the door whose number had come up. So let's say your initial choice is selected according to the uniform distribution: each of the doors has probability $1/3$ of being chosen. Now go back and look at S. You are initially going to choose door 2, say, with probability $1/3$. The only outcome in S with 2 in the first position is $(2, 3, 1, W)$. Therefore, we must assign $1/3$ as the probability of this outcome. Similarly, you are going to choose door 3 with probability $1/3$, and the outcome in S corresponding to this occurrence is $(3, 2, 1, W)$, and this outcome also is assigned probability $1/3$. Finally, you choose door 1 with probability $1/3$. But in S, when you choose door 1 you lose, and you can lose in two distinct ways. Our reasoning here only tells us that the event

you initially choose door $1 = \{(1,2,3,L),(1,3,2,L)\}$

has probability 1/3; without further assumptions the probabilities of the individual outcomes $(1,2,3,L)$ and $(1,3,2,L)$ are not uniquely determined. But for our problem these outcome probabilities are irrelevant, as we now see. For let $P(1,2,3,L) = a$, $P(1,3,2,L) = b$ where $a + b = 1/3$. The event we are interested in is

you win $= \{(2,3,1,W),(3,2,1,W)\}.$

From the rules,
$$P(\text{you win}) = 1/3 + 1/3 = 2/3.$$

Moreover,

$$P(\text{you lose}) = P(\text{you initially choose door 1}) = 1/3.$$

So this answers the question—according to our assumptions about the game, switching choices gives you probability 2/3 of winning and 1/3 of losing.

Now, what happens if you don't switch? To calculate this, let's start afresh with a new sample space based on not switching and simply go through all the possibilities as we just did above for the case of switching. We have a sample space S that can be written

$$S = \{(1,2,1,W),(1,3,1,W),(2,3,2,L),(3,2,3,L)\}.$$

The third component of each 4-tuple contains the same number as the first component since now you do not switch your choice. Using the same argument as in the case of switching, and assuming your initial choice of door is again dictated by a uniform distribution, we calculate

$$P(\text{you lose}) = P(2,3,2,L) + P(3,2,3,L) = 2/3,$$

and then $P(\text{you win})$ must be 1/3. Our conclusion is: switching gives you probability 2/3 of winning the car, and not switching gives you probability 1/3 of winning it. So you are twice as likely to win if you switch than if you don't.

Notice how we had to convert the original somewhat loose verbal description of the problem into a precise mathematical abstraction by making implicit assumptions apparent (i.e., each of your possible initial choices of door are equally likely). This is typical of many problems in probability—they are often phrased in an ambiguous way leading to more than one possible interpretation, and therefore resulting in several possible mathematical models. In Chapter 3 we will meet another version of the car and goat problem.

1.6   Exercises for Chapter 1

1. A single die is rolled, and then a coin is flipped twice. (a) Describe a sample space S giving all outcomes from this experiment. (b) Assume all outcomes in (a) have the same probability. Find the probability of the following events: 6 is rolled and at least one head turns up; an even number is rolled, and head turns up on the second toss; at least one head turns up, and a number less than 5 is rolled.

2. Consider the following variation of the car-goat problem. This time there are four doors, three goats, and one car. You choose a door at random and then the host selects a door with a goat behind it at random, which he opens. Suppose you switch to one of the other two doors, picking one at random. What is the probability now of winning the car? What is the probability of winning the car if you don't switch?

3. Suppose my alarm clock goes off sometime between 6 and 7 AM. Describe a sample space with each outcome a possible instant the alarm can ring. Is the sample space continuous or discrete?

4. Suppose in exercise 3 the alarm can only ring at 5 minute intervals: at 6 AM, 6:05 AM, 6:10 AM, etc. Now describe a sample space with each outcome a possible instant the alarm can ring. Explain why this sample space is discrete.

5. Assume that the discrete sample space in exercise 4 has the uniform distribution, and that I always have the same dream between 6:18 and 6:43 AM if I am sleeping. Suppose I will be sleeping during this time unless the alarm awakens me for the day. Find the probability of the alarm interrupting my dream.

2

How to Count: Birthdays and Lotteries

> Yet they, believe me, who await
> No gifts from chance, have conquered fate.
>   Matthew Arnold, *Resignation*

2.1   Counting your birthdays

There is a very famous problem about birthdays showing how the answers to certain problems can defy our intuition. The problem can be phrased this way: suppose you are at a party, in a hall filled with people. How many people do you think have to be present before the probability that at least two people have the same birthday is about 1/2? Having the same birthday here means the month and day must match; the year is irrelevant. Suppose, for example, there are 30 people at the party and someone comes over to you claiming at least two people there have the same birthday. You know he is a stranger to the group—he has no inside information. He wants to bet $10 that he is right. Is this a good bet to make? If you take this bet you will win only if everyone in the hall was born on a different day of the year. Since there are 365 days in the year and only 30 people present, it might seem quite likely that there are no repeats of birthdays in the place and that the $10 bet would be quite favorable to you.

Let's digress briefly to develop some ideas and language useful for talking about events. An event is just a set, and there is a standard way mathematicians discuss sets and build new ones from old ones by means of certain set operations. Suppose in a sample space S there are two events which we denote by A and B. A consists of some bunch of the outcomes and B consists of some other bunch of them. We now want to consider a new set (event) defined in terms of A and B, namely, the set of all the outcomes in S that are in either A or in B. This set will be expressed by the notation $A \cup B$, called the *union* of the sets A and B. This is the union operation. An outcome in the union is allowed to be in both A and B. If we consider, for example, the sample space S of all possible outcomes when you toss a coin twice and let the following sets be defined

$A = \{$the first toss shows head$\}$, $B = \{$the second toss shows head$\}$,

then we can write the sets in terms of their outcomes explicitly as

$$
\begin{aligned}
S &= \{(H,H),(H,T),(T,H),(T,T)\} \\
A &= \{(H,H),(H,T)\} \\
B &= \{(H,H),(T,H)\} \\
A \cup B &= \{(H,H),(H,T),(T,H)\}.
\end{aligned}
$$

In this example the outcome (H, H) happens to be an element in both A and B. In similar fashion, let us write $A \cap B$ for the set of all elements in A and also in B. In the above example $A \cap B = \{(H, H)\}$. This set is called the *intersection* of A and B. There is a third important operation which builds a new set using a single set rather than two sets as in union and intersection. Define A^c to be the set of all outcomes in the sample space S that are not in A, called the *complement* of the set A. In the above example $A^c = \{(T, H), (T, T)\}$. It is always true that $P(A) + P(A^c) = 1$; just use the rules for computing probabilities.

The notions of union and intersection can be extended to more than two sets in the obvious manner: if there is a collection of sets given, the union of the collection is the set of all outcomes in at least one of the sets of the collection, and the intersection of the collection is the set of all outcomes in all of the sets of the collection.

There is also an important way to show how two sets are related, written $A \subset B$. This is the *inclusion* relation, read "A is included in B," and means that every outcome in A is an outcome in B. For instance, both of the relations $A \cap B \subset A$ and $A \cap B \subset B$ are always true. Whenever $A \subset B$ it is always the case that $P(A) \leq P(B)$, since any outcome contributing to the probability of A also contributes to B. Figure 2.1 gives a Venn diagram illustrating the above operations.

Now let us consider the empty set, the set with no outcomes. At first, some people think this is just one of those useless ideas mathematicians think up to torture the rest of humanity. But the empty set is important if we want the above operations on sets always to produce a set. For example, consider rolling a pair of dice once. If A is the set of all outcomes giving the sum of the faces a number greater than or equal to 8, and B is the set of all outcomes with the first die showing 1, then $A \cap B$ has no outcomes

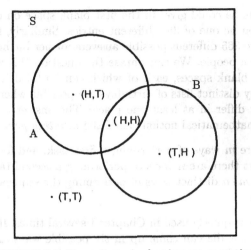

FIGURE 2.1. Venn diagram showing set operations

in it, and so would not even be a set if the empty set were to be shunned. So, embracing the empty set allows the above operations to have a very desirable property, what mathematicians call *closure*: you start with a set or sets and after you form a union, intersection, or complement you still have the same kind of object, a set. The empty set is also important from an epistemological point of view—it makes perfectly good sense, for instance, to refer to the set of all green-eyed winged unicorns in the Sahara desert without having to worry whether there really are such creatures.

Returning from this brief excursion into set theory, let us try to construct a sample space for the birthday problem. As usual, we must think carefully in order to construct an appropriate probability space, a reasonable abstraction of the problem. Let us generalize the problem and suppose there are r people in the hall, and let us simplify slightly by assuming only 365 possible birthdays—if someone was born on leap year day we will agree to give that person's birthday as March 1, say. Now, if $r > 365$ then at least two people must have the same birthday, so we will assume r less than or equal to 365. Suppose each person has been numbered from 1 to r, and that we have a list of r blank spaces as follows:

$$(--, --, --, \cdots).$$

We shall approach each person in the group, find out that individual's birthday, and enter it in the blank space corresponding to that individual's assigned number. We end up with a list telling us the birthday of each person in the hall; the 25th person, for instance, has birthday entered in the 25th blank line of the list. The sample space is the set of all the possible lists you could get in this way. How can one calculate this number of possible lists? When we go up to the first person and ask his birthday there are 365 possible answers he could give. In the first blank space on our list above, there could then be one of 365 different entries. Similarly, for the second person there are 365 different possible answers for her birthday, and so on for each of the r people. We can phrase the question this way: given the above list of r blank spaces, each of which can be filled in 365 different ways, how many distinct lists of birthdays are possible where two lists are distinct if they differ in at least one place? The answer to this problem depends on a mathematical notion called the *counting principle*:

> If there are m ways of performing a first task, and for each of those ways there are n ways of performing a second task, then there are $m \cdot n$ distinct ways of performing the sequence of the two tasks.

This principle was quietly used in Chapter 1 several times. If a pair of dice are rolled, the first die can come up in six possible ways and for each of those the second die can come up in six possible ways, so the counting principle shows there are $6 \cdot 6 = 36$ possible pairs of the first roll matched with the second roll. The counting principle is proved by enumerating all the possibilities—each one of the m possible ways of performing the first task can be matched with n possible ways of performing the second task, so if you add up all the distinct ways of matching you get $m \cdot n$; this is exactly how we counted the 36 distinct ways of rolling two dice. Now it is not hard to see the truth of the counting principle for any finite number of tasks, not just for two. If we have r tasks and the first can be done in m_1 ways, and for each one of those ways the second can be done in m_2 ways, and so on until the rth can be done (whatever the preceding choices) in m_r ways, then the totality of distinct ways of performing the sequence of all the tasks is the product $m = m_1 m_2 \cdots m_r$.

Now let's apply the counting principle to the present situation. Each time we fill in a birthday we are performing a task with 365 possible results. Once we have filled in any first bunch of blanks with birthdays, the next blank can be filled in 365 different ways. By the counting principle, the totality of distinct lists is $365 \cdot 365 \cdots$, r times, or 365^r. This is great—now we know what the sample space looks like. It consists of 365^r outcomes, each outcome being a list (or r-tuple) of r days of the year. For example, if $r = 3$, there are 365^3 different lists, and the list (March 3, January 20, June 6) would indicate that persons numbered 1, 2, and 3 have birthdays on March 3, January 20, and June 6, respectively. Before we ask the r people their birthdays we know only that exactly one of the lists in the sample space describes the true situation, but we don't know which list it is until we have the birthday information for all r people.

The next job is to decide on a probability distribution for the outcomes. Let us assume each list is equally likely, namely, that the distribution is uniform. Is this assumption reasonable for our mathematical model? Consider a civilization where the inhabitants are only fertile in the month of

June, and there is a nine month gestation period. In this civilization every-one would celebrate birthdays around February or March, certainly nobody would be born in August, say. The uniform distribution assumption for this society would lead to an inappropriate mathematical model. Instead, what is needed in this case is a distribution assigning probability zero to most of the lists, any containing August dates, for example, and concentrating the distribution mass on lists primarily in the February-March range. How would such a distribution be determined? We will return to this question later; again it relates to the important Law of Large Numbers. The essen-tial answer to the question is that we should estimate the probabilities of particular lists by studying the birth certificates of the society and looking at the relative frequencies of particular lists in many random selections of lists. In the ordinary human society we are part of, the uniform distribu-tion is indeed confirmed by such data to be a reasonable approximation to the true state of affairs. Therefore, we adopt a model with the uniform distribution—each list has probability 1 divided by 365^r, more easily writ-ten 365^{-r}.

So we are done in the sense that we have constructed what looks like a reasonable probability set-up: a sample space together with a probability distribution. The only thing left to do now is calculate the probability of events that we may be interested in, in particular the event "at least two people have the same birthday," which we call the event A. Now the uniform distribution has the pleasant property that to calculate the probability of an event all you have to do is *count* the number of outcomes in the event and divide by the total number of outcomes in S, the sample space; this follows since each outcome has the same probability, 1 divided by the total number of outcomes in S, and probabilities of an event are calculated by adding up probabilities of all outcomes in it. The problem has been boiled down to counting the number of outcomes in A. Fine, how do we actually do it? You can see this is a messy situation. Any time a list has at least two dates the same on it, it will be in A. There are an awful lot of those lists around and it doesn't seem easy to get a handle on how to count them. Not until we think of a little trick, that is.

Consider A^c, the complementary event to A. We can describe A^c as the totality of lists where all the birthdays are different. It *is* easy to count the outcomes in A^c: the first entry on such a list has 365 possible days, the second entry must be different from the first and so has 364 possible days, the third entry must differ from the first two and so has 363 possible days, and so forth. When we reach the last, rth entry, there are $365 - (r - 1)$ possible days since this last entry must differ from each of the preceding $r - 1$ days. By the counting principle, the total number of ways the list of r days can be filled out is $365 \cdot 364 \cdots (365 - (r - 1))$. According to the rules for calculating probabilities, we obtain

$$P(A^c) = \frac{365 \cdot 364 \cdots (365 - (r - 1))}{365^r}$$

$$= \left(1 - \frac{1}{365}\right) \cdot \left(1 - \frac{2}{365}\right) \cdots \left(1 - \frac{r-1}{365}\right).$$

The above formula can be evaluated with a calculator for any value of r. As r increases, we feel intuitively that it becomes less likely for all birthdays to differ since the number of people and hence the number of possibilities is increasing. This is indeed true; a calculation shows that as r increases the probability that all birthdays differ, $P(A^c)$, decreases. When $r = 23$ this probability dips below $1/2$ for the first time, and so $P(A)$ exceeds $1/2$ for the first time when $r = 23$. Put another way, if there are 23 people in a room the odds are in your favor that at least two people have the same birthday. Moreover, the more people there are the better the odds are in your favor, and they go up quickly: for $r = 30$, $P(A)$ is already approximately .7, and for 50 people it is .97. Most of us find it surprising on first hearing of this problem that so few people are necessary to get such high probabilities of a duplication of birthdays. So beware of the person who approached you at the beginning of this chapter with a \$10 bet; from your point of view it's a bad one.

2.2 Following your dreams in Lottoland

I have before me a New York State Lotto ticket for the Pick 6 game. The game is played this way: there is a panel of numbers from 1 to 54. The player marks 6 of these numbers. When the lottery drawing occurs, the player wins (at the first prize level) if all six numbers he chose match the drawn numbers. All the winners split the purse. The minimum play is two game panels for \$1. At the bottom of the ticket it says "Follow your dreams ... within your means." What we'd like to do now is investigate if, after following your dreams in Lottoland, you are likely to attain them, or will your dreams be more likely to be so far ahead of you that you will lose them (as well as all the money you gambled away).

The probability of winning is $1/25{,}827{,}165$ on one game panel; this is written on the ticket. This number is very small. One way we can feel its smallness is to compare it to an event whose probability we have some gut instincts about. Take a fair coin ("fair" means each side is assumed to have probability $1/2$ of turning up). Then it is less likely that you will win on one game panel than it is to obtain *24 consecutive heads in tossing the coin.* This hardly seems to be the stuff reasonable dreams are made of.

Let's see how to compute the probability they give on the ticket. This is an instructive exercise teaching us something about a bunch of things where the order in which they are written down is important and the contrary case where it is just the *set* of things given without the order being important. You must choose six numbers, so let's consider the set S of all 6-tuples (a, b, c, d, e, f), where each coordinate represents a distinct number from 1 to 54. How many such 6-tuples are there? The first entry can be written in 54 different ways, the second in 53 different ways (since the second must differ from the first), the third in 52 different ways, etc., until the sixth entry can be filled out in $54 - 5 = 49$ ways. The counting principle gives us a total of $54 \cdot 53 \cdot 52 \cdot 51 \cdot 50 \cdot 49$ ways of filling out this *ordered* list. Our counting procedure, for example, counts the lists $(21, 54, 1, 17, 8, 32)$ and $(1, 32, 8, 17, 54, 21)$ as different because we are counting ordered arrangements. But in the Lotto problem the order is irrelevant; it is only the set of numbers we write down, not the order in which we write them, that is important. So our total number of ways given above is bigger than we actually want because for any set of numbers it counts all the permutations (that is, orderings) of that set. Suppose we had chosen the numbers in the set $\{21, 54, 1, 17, 8, 32\}$. In how many ways could we have written these six numbers in an ordered 6-tuple? Think of six blank spaces. You can fill the first in six possible ways, the second in five ways, etc., so you have for each choice of numbers a total of $6 \cdot 5 \cdot 4 \cdot 3 \cdot 2 \cdot 1 = 720$ ways of writing them down.

If we have all the possible *ordered* arrangements, then the way to calculate all the *unordered* arrangements is to divide by 720. The number of these sets of six numbers is therefore

$$\frac{54 \cdot 53 \cdot 52 \cdot 51 \cdot 50 \cdot 49}{720} = 25,827,165.$$

A convenient sample space for this problem is the set of all these 25,827,165 outcomes, each one representing one of your possible selections on a game panel. The rules of the game are supposed to ensure the randomness of the drawing, that is to say, the agreement with the uniform distribution. A nice way to think of the Lotto drawing is this: one of the selections in your sample space is chosen at random by New York State. If it is the unique one (written in red ink, say) that was your choice, you win. But the probability of the event that New York State chooses this special 6-tuple is 1/25,827,165.

Of course, if you buy lots of tickets you increase your chances of winning. Recently a group tried to select all possible numbers in a Lotto game; the high value of the purse made it worthwhile for them to try and do this. Now, buying all the numbers creates practical problems and they failed to get them all, but they got enough so they did win. This idea of forming a company that tries to cover the field was thought by some to be illegal but the winnings were paid. It is very difficult for the typical individual Lotto player to unleash his fantasies adequately if he knows he is competing against such large groups with much capital. This could endanger the popularity of the games. It will be interesting to see whether laws are passed to prevent such groups from trying to corner the market in the future.

From the point of view of betting, then, Lotto is a pretty miserable game, and it would be better for you to follow your dreams by just going to sleep and saving your money. "But someone has to win," the avid player shouts, and that's true, but someone also has to get struck by lightning. Probability, in this case, is not claiming impossibility for your big win, only its extreme unlikelihood. If you want to play Lotto for the thrill of gambling, that's another story. Probability only talks about the rational realm, the chances of your success. And those chances in this game are quite slim.

2.3   Exercises for Chapter 2

1. Suppose Max is attending the convention for People Born in January. Max and seven others there get stuck in an elevator and, to pass the time until help arrives, Max tries to calculate the probability that all eight trapped people have different birthdays. What answer should Max get? What is the probability that at least two have the same birthday? (Assume all birthdays are in January.)

2. A party of six people goes to the theatre. In how many ways may they be seated in six adjacent seats?

3. Suppose three men and three women from a singles club are to be seated in the six seats of exercise 2 in such a way that two people of the same sex never sit next to each other. In how many ways can this be done?

4. Suppose I have seven adjacent spaces and I want to fill the spaces with the letters H and T such that I write the letter H four times and the letter T three times. How many distinguishable patterns can be formed in this way?

5. Suppose the Lotto game in Section 2.2 is altered by marking six numbers from a panel that now contains only the numbers from 1 to 40. Without doing any computation, tell whether the probability of winning on one game panel is more or less than the game given in Section 2.2, and explain your reasoning. Now calculate this probability explicitly.

3
Conditional Probability: From Kings to Prisoners

> The naked hulk alongside came,
> And the twain were casting dice;
> "The game is done! I've won! I've won!"
> Quoth she, and whistles thrice.
> Samuel Taylor Coleridge, *Rime of the Ancient Mariner*

3.1 Some probability rules. Conditional Probability

Again let's roll a pair of dice once, and let's consider the probability set-up with our now familiar 36 possible outcomes and the uniform distribution on the resulting sample space. Let A be the event "rolling a 7"; in Chapter 1 we saw that this event consists of six outcomes each having probability 1/36, so $P(A) = 6/36 = 1/6$. Let B be the event "the first (red) die comes up 1." If we write down all the outcomes in B, we find again six outcomes, namely, $(1,1),(1,2),(1,3),(1,4),(1,5),(1,6)$. So $P(B) = 1/6$ too. Now suppose we compute $P(A \cup B)$. By definition $A \cup B$ is the event occurring when either 7 is rolled or a 1 appears on the first die, or both happen. Write down all the outcomes described by this situation and you will find 11 outcomes: the six ways of rolling 7, which includes the outcome (1,6), plus the other five outcomes with 1 in the first position. According to our rules for computing probabilities $P(A \cup B) = 11/36$. How about $P(A \cap B)$? Well, $A \cap B$ is the event occurring when both A and B happen, namely, when the first die rolls 1 and we also get 7; this can happen in one way, when (1, 6) appears. Therefore $P(A \cap B) = 1/36$. Notice the validity of the following formula:

$$P(A \cup B) = P(A) + P(B) - P(A \cap B).$$

The reason why this formula holds is pretty easy to grasp. In order to calculate the left-hand side we have to add up the probabilities of all the outcomes in the union set. The union set consists of outcomes in A or B or both. On the right-hand side, $P(A)$ adds up for us all the probabilities of outcomes in A, and $P(B)$ adds up probabilities of all outcomes in B. But $P(A) + P(B)$ adds the probability of any outcome in *both* A and B exactly twice, once because it is in A and once because it is in B. So to get these outcomes in both A and B counted exactly once, we must subtract the term $P(A \cap B)$. The formula above is completely general; it holds for all probability spaces and all pairs of events A and B. If we have events A and B such that there are no outcomes in both A and B, that is, $A \cap B$ is the empty set, then since the empty set has probability 0 the formula simplifies to

$$P(A \cup B) = P(A) + P(B) \qquad \text{(if } A \cap B \text{ is empty).}$$

This formula is nice and simple. It says that if the two events have no common outcomes (in this case we say the events are *disjoint*), then the probabilities just add up to get the probability of the union. You can think of the events A and B as two non-overlapping pieces of land and the probabilities as their areas—the union of the two pieces is just the accumulated land with area just the sum of the two areas. If the pieces of land overlap, then the total property has less area than the sum of the individual areas and you have to use the first formula.

Let's go back to the probability space associated with rolling the pair of dice once and now let's suppose the event A is defined by "rolling a (total of) 6." The event A consists of the outcomes $(1,5), (2,4), (3,3), (4,2)$, and $(5,1)$, so $P(A) = 5/36$. We have already determined that B, the event "rolling 1 with the first die," has probability $1/6$. But now suppose you are given some new information. Suppose you are told that indeed it is the case that 1 was rolled with the first die. The question now is, *given this additional information, what is the value of $P(A)$?* It is natural to expect new information to alter your ideas of the uncertainties in the experiment, and therefore cause you to reevaluate the probabilities. In the above example, if it is known that 1 was rolled with the first die, the only possible outcomes are $(1,1), (1,2), (1,3), (1,4), (1,5), (1,6)$; we can ignore the other outcomes because they cannot occur based on the given information. We thus have a new sample space, and it is natural to suppose that in this new sample space each outcome is equally likely, since we started with an equally likely distribution. So *given* that 1 was rolled with the first die, the event A now consists of the unique outcome $(1, 5)$ and it seems as though its probability should be 1/6. The additional information updated the probability of A from 5/36 to 1/6.

In general, if we have a probability space and then new information arrives, it makes sense for us to update the probability space using the information, and therefore the computation of probabilities is updated based on this new information. When this sort of thing occurs, the updated probabilities are called *conditional probabilities* given the new information. Let us now define $P(A/B)$, the conditional probability of any event A given an event B, to be

$$P(A/B) = \frac{P(A \cap B)}{P(B)}.$$

This definition has an intuitive appeal. If B is known to occur, then any outcomes in the original sample space not in B cannot occur, so it makes sense to restrict the outcomes in A we are looking at only to those outcomes also in B; that is, $P(A/B)$ should be related to $P(A \cap B)$. In fact, we have defined it to be a value proportional to it. Since $P(B/B)$ should equal 1 on intuitive grounds, we see that the proportionality factor $P(B)$ on the right-hand side of the formula gives you the right answer to get a probability. In the above example, $A \cap B$ consists of the unique $(1, 5)$ with probability 1/36, and $P(B) = 6/36$, so the formula gives us $P(A/B) = (1/36)/(6/36) = 1/6$ as obtained above. The right side of the conditional probability formula always defines a probability distribution as long as $P(B) \neq 0$; we won't define conditional probabilities when the conditioning event has probability 0. Notice that by multiplying both sides of the conditional probability formula by $P(B)$ we obtain another useful version of the formula:

$$P(B) \cdot P(A/B) = P(A \cap B).$$

The updated conditional probability may have the same value as the original probability: if A is "rolling a 7" and B is " the first die comes up 1," then it is a simple exercise to obtain $P(A/B) = P(A) = 1/6$. In cases such as this where the conditional probability is the same as the original probability, we say that A is *independent* of B: intuitively, the added information provided by B did not give any new information about A in the sense that the conditional probability of A is no different from the original probability. This idea of independence is one of the major topics in probability theory. We start talking about it in earnest in Chapter 5.

3.2   Does the king have a sister?

Consider the following problem to test our skills with conditional probability (it appears as an exercise in [28]):

> The king comes from a family of two children. What is the probability that the other child is his sister?

The sample space for this problem can be considered to be the set S of four pairs $(B, B), (B, G), (G, B), (G, G)$, where B stands for "boy" and G stands for "girl" and the first and second positions in the pair denote first and second born children, respectively. To be able to do the problem some assumptions must be made. Once again, we shall assume each of the four outcomes is equally likely. Let U be the event "one child is a girl" and V be the event "one child is the king." What we want to calculate here is $P(U/V)$. Using the formula, we have

$$P(U/V) = \frac{P(U \cap V)}{P(V)} = \frac{P(\text{one child is } B \text{ and one is } G)}{P(V)} = \frac{2/4}{3/4} = 2/3.$$

This problem is tricky—a lot of people think the answer should be $1/2$ as in the car and goat problem. If the question had been "what is the probability that a person's sibling is a sister," then the answer would be $1/2$. But in the given problem you are sneakily given the information in the wording of the problem that one child, the king, is male, and that information eliminates the outcome (G, G) in the sample space as a possibility. The remaining three outcomes of S become the conditional, or updated, sample space of which two outcomes have a B and a G. This problem illustrates once again how careful you have to be when you are interpreting the information a problem is conveying. If there are ambiguities in the wording, different interpretations may lead you to radically different sample spaces and then to different answers.

3.3   The prisoner's dilemma

A good exercise to get you thinking carefully about conditional probability is a problem called *the prisoner's dilemma* (there is a completely different and famous problem related to game theory that goes under the same name; see, e.g. [4]). One version of this problem goes as follows: Consider three prisoners, A, B, and C. Two of the prisoners are to be released, and the prisoners know this, but not the identities of the two. Prisoner A asks the guard to tell him the identity of one prisoner other than himself who is to be released. The guard refuses and explains himself by saying to prisoner A, "your probability of being released is now $2/3$. If I tell you that B, say, is to be released, then you would be one of only two prisoners whose fate is unknown and your probability of release would consequently decrease to $1/2$. Since I don't want to hurt your chances for release I am not going to tell you." Is the guard correct in his reasoning?

The answer to this problem is not so obvious; it takes some analytical digging to find out why the guard's statement sounds a little too glib. The guard is thinking of the sample space of all possibilities of two prisoners released. This can be represented by the set of three outcomes

$\{A, B\}, \{A, C\}, \{B, C\}$, where curly braces are used to indicate unordered pairs, and the outcome represented by a pair is that those two prisoners are released. So far, the guard has constructed a fine sample space. To get a probability space we use a uniform distribution, which means the parole board chose the prisoners to be released at random. (An assumption like this is necessary for many of these problems. Otherwise there is no obvious probability distribution and the problem can't be done until you get some distribution.) Each of the above outcomes is therefore assigned probability $1/3$, and the guard's first statement about the probability of A's release being initially $2/3$ is correct. The trouble begins when the guard seems to be saying

$$P(A \text{ is released/guard says } B \text{ is released}) = 1/2.$$

The first thing to notice is that the conditional probability given above cannot be computed in terms of the given sample space the guard defined: we simply do not have the event "guard says B is released" to condition by. This suggests the need for a more complex sample space incorporating the guard's statement. Consider the sample space given by the following four outcomes

$$O_1 = \{A, B, \text{guard says } B \text{ is released}\},$$
$$O_2 = \{A, C, \text{guard says } C \text{ is released}\},$$
$$O_3 = \{B, C, \text{guard says } B \text{ is released}\},$$
$$O_4 = \{B, C, \text{guard says } C \text{ is released}\}.$$

These give all the possibilities matching the release of two prisoners with a compatible statement by the guard. The event O_1 is equivalent to the event that A and B are released (the guard has no choice of statement), so has probability $1/3$, and similarly $P(O_2) = 1/3$. Now matters get really interesting. Since the union of O_3 and O_4 is the event $\{B, C\}$, this union has probability $1/3$. But without some further information there is no way to determine the individual probabilities of O_3 and O_4. Usually, one assumes each of these events is equally likely with probability $1/6$; this corresponds to the guard tossing a coin in case both B and C are released to determine which of the two he should identify to A in his statement. However, he could certainly use some other procedure, for instance, always identifying B.

First we will take each event with probability $1/6$. In this case

$$P(A \text{ is released/guard says } B \text{ is released})$$
$$= \frac{P(O_1)}{P(\text{guard says } B \text{ is released})} = \frac{1/3}{1/3 + 1/6} = 2/3,$$

proving the conditional probability of release of A the same as the original unconditional probability. (The term $1/3 + 1/6 = 1/2$ in the formula appears because the event "guard says B is released" is the union of the events O_1 and O_3, which are disjoint. To get the probability of the union, add up the individual probabilities in accordance with the second formula of Section 3.1.) The same argument works in case the event is "guard says C is released," and again we get 2/3. So we have solved the problem: the guard has not changed the probability of A's release by giving his statement, that is, the event "A is released" is independent of the event "guard says B is released."

But now consider the case where the guard *always* identifies B in his statement when both B and C are to be released. Then O_3 has probability 1/3 and O_4 has probability 0. In this case, the term $1/3 + 1/6$ in the formula above becomes $1/3 + 1/3 = 2/3$, so the conditional probability is $(1/3)/(2/3) = 1/2$ and now the conditional probability *is* 1/2, as the guard had said. So the guard *can* change the value of the conditional probability by altering the way in which he determines his statement when he has a choice (when B and C are both released). He could choose to identify B with any probability from 0 to 1/3, and the conditional probability will then be some number between 1/2 and 1. Does this really mean the guard has control over A's fate, as he believes, by the way he determines his statement? This goes against our intuition. If this were true, then if the guard simply whispered his statement to himself in private rather than telling A, wouldn't the same argument given above show that in this way too the guard can alter A's fate? The decision on which prisoners to release was, after all, made by a parole board and had nothing to do with the guard's statement. What this seems to be suggesting to us is that we should start out by *assuming* the independence of the event "A is released" from the guard's statement. If this is done, then the conditional probability in the formula above is 2/3 and the only way this can happen is if O_3 and O_4 each are given probability 1/6, as was done for the first solution. So the first solution is the one that meshes with the real world as we perceive it. The other solutions, while mathematically correct, don't correspond to the model we need here. Observe that if we are interested in the *unconditional* probability $P(A$ is released$)$ rather than a conditional probability, then the answer is just $P(O_1) + P(O_2) = 2/3$ regardless of the way the guard chooses to make his statement.

We have just seen that the solution giving independence of the event "A is released" from the guard's statement is the reasonable one for the prisoner's dilemma problem. Let's see how another problem leads to the same mathematical model as the prisoner's dilemma problem, except that now *any* possible solution turns out to give a reasonable real-world interpretation. We are going to consider a version of the car and goat problem different from the one discussed in Chapter 1. There, you recall, we assumed your choice of door was random and the car was behind door 1. The original statement of the car and goat problem in the newspaper was a little fuzzy; you could interpret the wording in several ways and get several different problems. The most common version of the problem was described in

Chapter 1. Another version, analyzed by Gillman [13], assumes you always choose door 1 initially, but the car and goats are distributed at random (namely, uniformly) behind the doors. The host opens either door 2 or 3. You then switch doors. The problem is: find the conditional probability that the car is behind door 2, given that the host opens door 3. This will give the conditional probability of winning if you switch. Gillman showed that the answer depends on the conditional probability of the host opening door 3, given that the car is behind door 1. If the events "winning the car" and "opening door 3" are identified with the events "A is released" and "guard says B is released," respectively, in the prisoner's dilemma, we find the problems essentially equivalent. The conditional probability of winning the car given that the host opens door 3 varies between 1/2 and 1, just as the conditional probability of A's release given that the guard says B is released varies depending upon the probability of the guard making his statement. A major difference between the problems, however, is in the mathematical model the real-life situation suggests. In the prisoner's dilemma, as we have mentioned, we feel at the outset that A's release should be independent of any statement the guard makes, and we then should build our sample space to reflect that fact. In the case of this new version of the car and goat problem, the nature of the events considered now suggests it is reasonable to assume dependence, so we get a sensible model by assuming any one of the possible solutions discarded previously. That means the answer can reasonably be any number between 1/2 and 1.

3.4 All about urns

Suppose we have an urn containing ten balls, six red and four black. The balls are of the same size and have been mixed up well. You now choose a ball at random from the urn and mark down its color. You do not replace the ball and, after making sure the balls are again well mixed, you choose a second ball from the urn and note its color. Define the events

$A_1 = $ {first ball is red}, $A_2 = $ {second ball is red}.

We are interested in the probabilities of the events $A_1, A_1 \cap A_2$, and A_2.

The probability of A_1 is easy to find. The mixing of the balls is a code expression common in such problems to mean "assume a uniform distribution." So the sample space can be represented by a set with ten outcomes, each one standing for a ball of a certain color with probability .1 of being selected. The event A_1 contains six outcomes, so its probability is .6. Now consider the event $A_1 \cap A_2$. Use the conditional probability formula in the form

$$P(A_1 \cap A_2) = P(A_1) \cdot P(A_2/A_1) = 6/10 \cdot 5/9 = 1/3.$$

The conditional probability in the formula equals 5/9 because the composition of the urn just before the second selection consists of five red balls and four black ones, and again we use the uniform distribution because the balls are once again well mixed. Finally, let's consider A_2. This event does not depend upon what happens in the first selection. But to calculate this probability we must consider all possibilities of what happened at the first selection because they all contribute a little probability "weight" to the event we're interested in. The important fact is

$$A_2 = (A_1 \cap A_2) \cup (A_1^c \cap A_2)$$

which just says that if we got a red ball on the second selection we either had a red ball on the first selection or *not* a red ball (i.e., a black ball) on the first selection. Since the events in parentheses above are disjoint (you can't simultaneously get both a red and black ball on the first selection), the formula in Section 3.1 for probabilities of disjoint unions gives

$$P(A_2) = P(A_1 \cap A_2) + P(A_1^c \cap A_2).$$

The first term of the sum was just calculated above—we got $1/3$. We calculate the second term in exactly the same way, using the formula

$$P(A_1^c \cap A_2) = P(A_1^c) \cdot P(A_2/A_1^c) = 4/10 \cdot 6/9 = 4/15.$$

[There are four black balls before the first selection, so the probability of the first selection yielding a black ball is $4/10$; then there are nine balls in the urn, six of them red, and the (conditional) probability of the second selection giving a red ball is $6/9$.] So $P(A_2) = 1/3 + 4/15 = .6$.

One of the interesting aspects of the urn problem is that we calculate the probability of an intersection of two events in terms of a given conditional probability. That is in contrast to the problems given in the previous sections where we calculated a conditional probability in terms of initially given probabilities of intersections. In the urn problem the naturally occurring probability is a conditional probability: the probability of the second selection depends on what happened on the first selection.

A second interesting fact is that A_1 and A_2 have the same probability, .6. This is not just a coincidence of the numbers chosen here. If we take a red balls and b black balls, and only assume there are at least two balls in the urn (so two selections are possible), an easy exercise in algebra shows A_1 and A_2 still have the same probability which, in this general case, is $a/(a + b)$. At first, this phenomenon may seem curious because the second selection takes place later in time than the first selection and this masks an important symmetry. Instead of thinking of choosing the balls in sequence, imagine that we simply choose *once*, reaching in simultaneously with both hands and selecting a ball with each. We can represent the outcome as an ordered pair like (R, B) where the first coordinate is the color of the ball in the left hand and the second coordinate the color of the ball in the right hand. The ball in the left hand could also be identified with a first ball chosen, and the ball in the right hand with a second ball. Now the choice of the ordered pairs leads to a uniform probability space in the same way as the descriptions of selections in Chapter 2. A symmetry is now apparent in this model: any ordered pair with R as the first coordinate corresponds to an ordered pair with R in the second coordinate by interchanging the first and second coordinates. The number of outcomes in the events A_1 and A_2 are therefore the same, and so we should not now find it surprising that they have the same probability. This way of thinking about the problem also shows that we should expect the same phenomenon at the nth selection if there are enough balls in the urn to make n selections, that is, the probability of selecting a red ball at selection n is still $a/(a+b)$. This example also shows you how looking at a problem in a slightly different way may give insights not so easily obtained from another perspective.

The above model of selecting balls without replacement can be altered in various ways. We could, for example, choose a ball from the urn and if it is red replace the ball and place one additional red ball in the urn. If the ball chosen was black, replace the ball and add one additional black ball to the urn. In this model, the total number of balls in the urn is increasing rather than decreasing. This is a particular case of the so-called *Pólya urn scheme* which provides a crude model of a contagious disease. Each selection of a ball represents sampling an individual in a certain population. The red ball means the person is infected with the disease; the black ball means that she is free of it. Each discovered infection indicates an increase in the probability of finding another infected individual, and each discovered healthy person indicates an increase in probability for finding another healthy person. Using refined versions of this model, the long-term course of the disease can be studied.

3.5   Exercises for Chapter 3

1. Let S be the usual probability space for a pair of dice rolled once, using the uniform distribution. Suppose A is the event "first die is odd," and B the event "second die is even." In words, describe each of the following events and calculate their probabilities. (a) $A \cap B$, (b) $(A \cap B) \cup (A^c \cap B^c)$, (c) A^c, (d) $(A \cup B)^c$.

2. Roll a pair of dice once. What is the probability of getting 11? What is the probability of getting 11, given that the sum of the faces is an odd number? What is the probability of getting 11, given that the sum of the faces is an odd number greater than 3?

3. Toss a coin four times. Find the probability of getting at least two heads. Find the probability of getting at least two heads, given that there was at least one head. Find the probability of getting heads on all four tosses, given at least two heads.

4. An urn contains five red balls and five black ones. A ball is chosen at random and then it, as well as another ball of the same color, is returned to the urn. A second ball is then chosen at random. Find the probabilities that (a) the first ball is red and the second is red, (b) the first ball is red and the second black, (c) the second ball is red, and (d) the second ball is black.

5. Let A, B, and C be any events such that A and $A \cap B$ have positive probability. Use the conditional probability formula to prove the relation
$$P(A)P(B/A)P(C/A \cap B) = P(A \cap B \cap C).$$

4

The Formula of Thomas Bayes and Other Matters

He proves by algebra that Shakespeare's ghost is Hamlet's grand-father.

James Joyce, *Ulysses*

4.1 On blood tests and Bayes's formula

There is a blood test for the HIV virus causing AIDS. This test is quite good in the sense that if an individual has the virus the probability of detection is high. How is such a probability estimated? As we have mentioned before, the Law of Large Numbers soon to be discussed justifies our intuitive notion that a probability can be estimated by considering relative frequencies. So in this case we can give the test to a large population where we know the disease, and therefore the virus, is present. If the test is positive for, say, 95 percent of this population, we can say that .95 is a rough estimate for what is called the *sensitivity* of the test, defined as

P(test is positive/disease is present).

Unfortunately, all medical tests give occasional false results. There are two villains here:

P(test is negative/disease is present)   (false negative)
P(test is positive/disease is absent)   (false positive).

If it is assumed that the test identifies the virus when it is present 95 percent of the time, then it follows from the rules of probability that the probability of a false negative is 5 percent, $1 - $ (the sensitivity). But how about a false positive? To get a handle on that, we need to do more estimating. Now we would need another population, a group of people who we believe to be free of the virus. The probability of a false positive is estimated by considering the relative frequency of those members of this population who test positive. (We are simplifying the estimation procedure here. Some of those we thought to be virus-free could later come down with the disease. The actual design of the experiment is more sophisticated than my description and takes account of this possibility.) If the test is reasonable, the probability of a false positive will be low, and ideally it would be nice to have a test where the probabilities of both false negatives and false positives are as small as possible.

Several years ago, it was suggested that all couples applying for a marriage license should be required to take the blood test for AIDS. It was argued that such a requirement could be very helpful in slowing the spread of the disease. Many experts, however, argued against this proposal as a waste of money and resources better used elsewhere. The blood test requirement for the AIDS virus was never implemented. Should it have been?

An English theologian and mathematician, Thomas Bayes (1702–1761), helps us analyze this problem. Let A and B be any events in a probability space. Then one version of a relation appearing in a posthumous article of Bayes is the following:

(Bayes's Formula)

$$P(A/B) = \frac{P(B/A) \cdot P(A)}{P(B/A)P(A) + P(B/A^c)P(A^c)}.$$

Bayes's formula can be checked very easily using the definitions and rules in Chapter 3. The left-hand side of the formula is just

$$\frac{P(A \cap B)}{P(B)}.$$

But

$$P(A \cap B) = P(B/A)P(A)$$

and

$$P(B) = P(B \cap A) + P(B \cap A^c) = P(B/A) \cdot P(A) + P(B/A^c)P(A^c).$$

Algebraic substitution gives you Bayes's formula. Now you may ask, "What is the big deal about this formula? Isn't it merely a slightly different form of something we already know?" Well, the algebra is indeed simple, but the formula contains a very important idea. On the left side, notice that the conditioning event, the one we are given, is B. On the right side, however, the conditioning events are A and A^c. So the formula tells us that if we are given probabilities conditioned on the events A and A^c, we will be able to calculate a probability conditioned on B.

Let us see how this works on the type of problem we have considered above. Suppose for a hypothetical disease the probabilities of a false positive and a false negative have both been estimated to be about 5 percent. Moreover, assume estimates show the disease appears in about 0.3 percent of the population. The question is to calculate the probability that the person has the disease given the test is positive. Define the events

$$A = \{\text{tested person has disease}\}, \ B = \{\text{test result is positive}\}.$$

Use Bayes's formula: the left side of the formula is precisely the quantity we want to calculate. For $P(B/A)$ we use the estimate of the sensitivity, $.95 = 1 - .05$, and for $P(A)$, the estimate .003 of the frequency of the disease in the population. Now $P(B/A^c)$ is the probability of a positive test result for the individual given that he does not have the disease; for this we use the estimate for the probability of a false positive, .05. Putting these numbers into Bayes's formula gives

$$P(A/B) = \frac{(.95)(.003)}{(.95)(.003) + (.05)(.997)} \approx .05$$

where the two wiggly lines mean "approximately equal to." You may find this result surprising. Interpreting probabilities as relative frequencies, we see that only about 5 percent of the time does a positive test result indicate a person who has the disease. Ninety-five percent of the time the positive test result incorrectly labels healthy people as diseased. A careful look at the algebra shows the following: the smaller $P(A)$, the frequency of the disease, the less reliable a positive test is in correctly identifying diseased individuals, and the larger $P(A)$ is, the more reliable the positive test becomes. So the positive test is going to be unreliable for relatively rare diseases even though the probabilities of false positives and of false negatives are both small.

Let us apply the above analysis to the test for the HIV virus. The blood test for HIV has acceptably small probabilities of both false positives and false negatives; data indicate that we may assume both of these probabilities are smaller than .1. Moreover, AIDS is a rare disease in the general population, with an estimated frequency of about 0.006 (using a 1988 estimate). Since individuals applying for a marriage license would not constitute a high-risk group, we can expect the frequency of AIDS in such individuals to be roughly the same low value as in the general population. The argument above, therefore, indicates that we can expect a positive blood test obtained from a marriage license applicant to be wrong most of the time. The small number of true AIDS cases the test might uncover in this low-risk population would not justify the enormous expense in resources nor the psychological turmoil of the victims of an incorrect identification. According to our conclusions above, it would make much more sense to apply our resources to high-risk groups of individuals since a positive test then becomes a more reliable indicator. To see this concretely, go back to the hypothetical disease discussed above. Assume the probabilities of false positives and false negatives are the same, but now suppose $P(A)$ is .1 rather than .003, namely, the disease occurs in about 10 percent of the population rather than in 0.3 percent. Bayes's formula now gives the value

$$\frac{(.95)(.1)}{(.95)(.1) + (.05)(.9)} \approx .68$$

as the probability of a diseased individual given a positive test. Now a positive test errs only about 32 percent of the time rather than 95 percent.

From the above discussion, we conclude that the decision not to require across-the-board testing for individuals applying for a marriage license was a wise one. This is a good example of how mathematics, in particular the ideas of probability, can uncover flaws in what at first may appear to be a reasonable course of action.

4.2   An urn problem

An important interpretation of Bayes's formula is that under certain circumstances a probability $P(A)$ (on the right-hand side of the formula) can be updated to a conditional probability $P(A/B)$, given evidence B (on the left-hand side). To get more insight into this point of view, let's consider an interesting problem similar to one discussed by Laplace on page 18 of [23]. An urn contains two balls, each of which can be white or black. We will select balls repeatedly from the urn with replacement according to the following procedure: mix well, select a ball, note its color, replace ball in urn, mix well, select a ball, and so on. Suppose the first two selections yield white balls. Find the probability of a white ball at the third selection.

To answer this question, we will postulate a *prior* distribution for the color of the balls in the urn, that is, a distribution assumed *before* the first two drawings are observed to give white balls. Let us suppose this prior distribution is random, with each ball white or black with probability 1/2. This means the composition of the urn will have both balls white with probability 1/4, both balls black with probability 1/4, and balls of differing colors with probability 1/2. Define events as follows:

$$D \;=\; \text{the balls have different colors.}$$
$$W \;=\; \text{the balls are both white.}$$

$$B \ = \ \text{the balls are both black.}$$
$$W2 \ = \ \text{first two selections yield white balls.}$$
$$C \ = \ \text{the ball is white at the third selection.}$$

We can answer the question easily using the rules of conditional probability. Since $P(C/B)$ and $P(W2/B)$ are both 0, we have

$$P(C \cap W2) \ = \ P(D)P(C \cap W2/D) + P(W)P(C \cap W2/W)$$
$$= \ (1/2)(1/8) + (1/4)(1) = 5/16$$

and

$$P(W2) \ = \ P(D)P(W2/D) + P(W)P(W2/W)$$
$$= \ (1/2)(1/4) + (1/4)(1) = 3/8,$$

so $P(C/W2) = (5/16)/(3/8) = 5/6$. Let's solve this problem a slightly different (and longer) way to get additional insight. We will calculate the updated conditional distribution of the balls in the urn given the observations of the first two selections. To calculate these updated probabilities, use Bayes's formula, which can be written as

$$P(D/W2) = \frac{P(W2/D)P(D)}{P(W2/D)P(D) + P(W2/W)P(W) + P(W2/B)P(B)}.$$

Since there are three possible *a priori* compositions of the urn, notice that Bayes's formula now has three terms in the bottom of the fraction rather than two. All the terms on the right-hand side of the formula can be calculated from the conditions of the problem. As we have seen above, $P(D) = 1/2, P(W) = P(B) = 1/4$, and $P(W2/D) = 1/4, P(W2/W) = 1, P(W2/B) = 0$. Substitute these values into Bayes's formula to get

$$P(D/W2) = \frac{(1/8)}{(3/8)} = \frac{1}{3}.$$

Since $P(B/W2) = 0$, we must have $P(W/W2) = 2/3$. We now have updated probabilities of the composition of the urn, based on the evidence of the two draws. Now we need only compute the probability of drawing a single ball (the third selection) using this updated distribution (which we denote by P^*):

$$P^*(C) \ = \ P^*(D)P^*(C/D) + P^*(W)P^*(C/W)$$
$$= \ (1/3)(1/2) + (2/3)(1) = 5/6,$$

the same answer as before. In this second method, we explicitly use the first two observations to give us a new conditional distribution; the first method gets the answer efficiently by using all the information at once on the prior distribution.

Laplace's problem differs from the one just given only in the choice of the prior distribution. Instead of assuming each ball has an initial probability of being white or black with probability 1/2, Laplace supposes each of the events D, W, and B have the same initial probability 1/3. By substituting these numbers into the above formulas, you can see that Laplace's answer to the problem is 9/10 rather than 5/6.

4.3   Laplace's law of succession

How much would you be willing to bet that the sun will rise tomorrow, given that it has risen each day for the past 5,000 years? Laplace, in [23], uses an urn model based on the drawing scheme of the preceding section to offer an answer to this question. Laplace's reasoning goes like this: the urn contains a large number of black and white balls; each trial (drawing) represents a day. The selection of a white ball from the urn corresponds to the sun rising. Just as we calculated the probability of choosing a third white ball from the urn, given that two white balls have already been chosen, it will be possible to calculate the probability of the next ball being white (sun rising) in an urn, given that the preceding 1,826,213 selections have resulted in white balls (the sun has risen about 5,000 years). Such a calculation, though, depends upon our assumption of a prior distribution for the composition of the urn (in the preceding section we assumed each ball could be black or white with equal probability). Since we are ignorant about such a distribution, Laplace assumes that all possible compositions of the urn are roughly equally likely. He does this by supposing a large number $N + 1$ of urns with urn i containing i white and $N - i$ black balls. Select an urn at random, and select n balls within this urn using the procedure of the previous section (selection with replacement). Given that these n balls turn out to be white, Laplace estimates the probability of the next ball being white at approximately $(N + 1)/(N + 2)$ for large N (we omit the details). This is called Laplace's law of succession. From this he concludes that a bet of 1,826,214 to 1 of the sun rising again tomorrow is reasonable.

From a modern point of view, there is a lot wrong with Laplace's model. One major problem is the justification of identifying an occurrence of an astronomical event with drawing a ball from an urn. Even if this were legitimate, another problem concerns the assumption that all compositions of the urn are equally likely simply because we are totally ignorant about the distribution of the compositions. The equivalence of ignorance with the notion of equally likely options was fairly common in the early days of probability; it has been referred to in the literature of the subject as the *principle of indifference*. This idea is suspect on an intuitive level (how can you say anything at all if you have no information), and leads to problems. For example, consider the three events $A =$ no rain tomorrow, $B =$ rain tomorrow stopping before noon, and $C =$ rain tomorrow not stopping before noon. If we treat these events as equally likely because of ignorance about rain, time periods, etc., we must assign the common value 1/3 as the probabilities of A, B, and C. On the other hand, we can just as readily consider the events A of no rain tomorrow and D = rain tomorrow sometime during the day. Now, ignorance translated into equal probabilities gives 1/2 as the probabilities of A and D. Similarly, it is easy to consider A^c as the union of a set of $N - 1$ disjoint events about which we are totally ignorant, thereby giving A the probability $1/N$. So we can get any probability for A that we please, just by being ignorant of enough things! Most modern probabilists consider it wrongheaded to equate ignorance with an equally likely distribution. Any distribution assumption used in applications should be based on knowledge, not ignorance. From time to time, however, the principle of indifference still crops up when people are desperate for a probability distribution but have no hard information leading to an appropriate one (see, for example, Section 4.5).

It is not clear whether Laplace was serious in his attempt to figure out the odds on the rising of the sun. Perhaps the figure of 1,826,214 to 1 was meant to be a little facetious in light of the eighteenth century's view of God as ultimate clockmaker, guardian of the regularity of natural laws. Indeed, right after Laplace gives the 1,826,214 to 1 odds, he says, "But this number is incomparably greater for him who, recognizing in the totality of phenomena the principal regulator of days and seasons, sees that nothing at the present moment can arrest the course of it" (i.e., the rising of the sun). So the pious person would be willing to risk much more on this bet. Laplace here is really talking about subjective probability, the topic we now turn to.

4.4 Subjective probability

The events we have been considering, you recall, model the outcomes of a repeatable experiment. Rolling dice, tossing coins, choosing someone from a population who may be infected with a virus— these are all experiments of the repeatable type. For events of this type, like "rolling a 7," the notion of relative frequency makes sense. For example, what is the relative frequency, or proportion, of getting a 7 in many repetitions of rolling a pair of dice? The Law of Large Numbers shows that under our usual assumptions the relative frequency in this case tends to stabilize close to 1/6, the probability of rolling a 7. So if the dice are rolled a million times and the total number of times 7 appears is divided by a million, we should get a number pretty close to 1/6, and the larger the number of repetitions taken the better this approximation should be. Now many probabilists just like to think about this repeatable type of event where relative frequencies get close to the probability. Most classical and modern probability research is about events of this type, and most standard treatments, like this one, stick principally to these repeatable events. Then you can just accept the various standard rules or axioms of the theory, some of which we have given in preceding chapters. From these rules, the Law of Large Numbers follows and then so does the connection between probability and relative frequency.

Although the classical theory of probability requires events that can be repeated, there are certainly events for which the notion of probability seems reasonable and yet these events cannot consist of outcomes from a repeatable experiment. In Chapter 1 the example of a jury trial was given with the event "the defendant is guilty." Such an event does not fit into a framework of repetition, but still one may want to attach a probability to such an event measuring the degree of belief one holds that the proposition is true. Perhaps you, a juror, feel 90 percent sure that the defendant is guilty. Such a probability would be a personal, or subjective, probability, different for each juror perhaps, and subject to revisions by further evidence. This interpretation of probability could still be subject to many of the rules of probability we have accepted for repeatable events; for example, the sum of the probabilities of the events "the defendant is innocent" and "the defendant is guilty" is 1. But now the Law of Large Numbers is meaningless; what takes its place is the idea that personal probabilities get close to one another with increasing evidence.

The concept of probability not as an immutable number belonging to an event but rather as a varying value depending on individual assessment makes many mathematicians feel uncomfortable with subjective probability. People who believe that the repeatable events are the only kind of events probability theory should consider are often called "frequentists" because they place all their faith in relative frequencies. On the other hand, there are those, the "subjectivists," who look at probability only from the subjective viewpoint, even for repeatable events. The subjectivists find in Bayes's formula the expression of the basic idea behind their beliefs; for that reason they and their results are often called *Bayesian*. Take a look at Bayes's formula again, and notice how on the right side $P(A)$ appears and on the left the conditional probability of A given B. The Bayesian views $P(A)$ on the right as an original or prior subjective probability, and the conditional probability on the left as the updated, or posterior, version of this subjective probability using the additional information supplied by B. The main criticisms of the Bayesian approach concern the need for assuming the prior probability distribution and the method of determining it. On the other hand, those who want to stick with repeatable events and relative frequencies can also be criticized for unnecessarily restricting the notion of probability. The debate between the frequency people and the Bayesians gets especially fiery in their approach to statistics, about which we'll speak more in Chapter 15.

4.5 Questions of paternity

Here is a problem in subjective probability: a man accused in a paternity case is found to have a genetic marker appearing in 1 percent of the adult male population. This genetic marker is found in the child and could only be transmitted to the child through his father, with the child 100 percent certain of acquiring the marker if the father has it. The question is to determine the probability that the man is the father, given that the child has the marker. Define the events

$$A = \{\text{the man is the father}\}, \quad B = \{\text{the child has the marker}\}.$$

Let us try to use Bayes's formula to calculate $P(A/B)$. Since the father always transmits his marker, $P(B/A) = 1$. Morever, $P(B/A^c) = .01$, because if the accused man is not the father, we can suppose the appearance of the marker in the child is equivalent to the appearance of the marker in the adult male population. Now comes the controversial part. To use Bayes's formula, we must have a prior estimate or guess for $P(A)$. Take this guess equal to .5, and put all the numbers in the formula to get

$$\frac{(1)(.5)}{(1)(.5) + (.01)(.5)} \approx .99.$$

This result is interpreted to mean that if one initially assumes the man's probability is .5 of being the father, then the updated probability that he is the father given that the child has the marker is about .99. Suppose, on the other hand, we had assumed a prior value of .001 rather than .5 for $P(A)$, and substitute into the formula again using this new value. Then

$$\frac{(1)(.001)}{(1)(.001) + (.01)(.999)} \approx .09,$$

and the posterior probability of the man's guilt sinks from .99 to .09. The crucial importance of the prior probability is clear from this example.

An interesting legal case involving these issues was reported in [31]. The case was similar to the hypothetical problem described above. Although the statistical methods used were not apparent from the article, it seems likely that Bayes's formula was used with a prior probability of .5 for $P(A)$ assumed in a probability analysis for the prosecution. The defendant had then been convicted based at least partly on the very large posterior probability that he was the father. This conviction was, however, overturned on appeal because of this probability analysis, which, in effect, assumed that the man committed the crime with a high degree of probability (.5) to prove that he committed it with an even higher degree of probability. This is an example of how the use of probability or statistical methods in a courtroom proceeding can backfire when used inappropriately. The prior assumption of .5 for $P(A)$ was only chosen because of the investigator's ignorance of the true value. The idea was: either the guy is guilty or not, and since I don't know I'll assign each possibility the probability .5. So here again is the principle of indifference at work, that bad habit of equating ignorance with equal probabilities. As we saw earlier, there's no justification for this practice; an assignment of a probability distribution should be based on positive knowledge of the model, not total ignorance. There is a fundamental difficulty with using Bayes's method in legal cases since any prior distribution assigning a positive probability of guilt can always be attacked as violating a person's right to be considered completely innocent until proven guilty. Incidentally, an interesting recent book about the uses of probability and statistical methods in the law is [10].

4.6   Exercises for Chapter 4

1. Suppose each of two balls in an urn can be either red, black, or green with probability 1/3. You choose a ball at random from the urn, note that it is green, replace the ball, and then choose once again at random. What is the probability that the second ball chosen is red? Answer the question for each of the other colors.

2. Roll a pair of fair dice once. Let A be the event "at least one die shows 6" and B the event " the sum of the faces gives an odd number." (a) Find $P(A/B)$ (b) Using Bayes's formula and part (a), find $P(B/A)$.

3. Suppose I have faith in my local weather reporter who says that the chance of rain tomorrow is 80 percent. Moreover, my friend, the seasoned sailor, tells me that whenever it will rain the next day, the type of cloud and appearance of the sky visible this evening only appears about 10 percent of the time. Whenever it will be clear in a 24 hour period, he adds, the present cloud and sky pattern occurs 60 percent of the time the evening before. Assuming I trust my friend's judgment, what would be my belief in the chance of rain tomorrow, given the appearance of the sky this evening?

4. A patient suffers from a condition which is fatal 50 percent of the time. One of the possible treatments for this condition involves surgery. Research has shown that 40 percent of survivors had surgery, and 10 percent of non-survivors had surgery. Find the probability of the patient surviving the condition if he has surgery.

5. (*Return of the car and the goats.*) You are playing the car-goat game as described in Chapter 1 with only one change: when the master of ceremonies asks whether you wish to switch your choice of door you toss a fair coin. If the coin falls heads you switch; if it falls tails you don't switch. Now suppose you win the car. What is the probability that you switched doors?

5

The Idea of Independence, with Applications

> To him, therefore, the succession to the Norland estate was not
> so really important as to his sisters; for their fortune, indepen-
> dent of what might arise to them from their father's inheriting
> that property, could be but small.
>
> Jane Austen, *Sense and Sensibility*

5.1   Independence of events

In the real world, we frequently encounter pairs of events such that the occurrence of one, we feel, has no bearing or influence on the occurrence of the other. For example, suppose I toss a coin once and note what comes up. Suppose then I repeat the procedure, giving the coin another toss and again note what comes up. Consider the events

$$H_1 = \{\text{head on toss 1}\}, \qquad H_2 = \{\text{head on toss 2}\}.$$

Under most circumstances most people would have the strong intuitive feeling that the occurrence of H_1 gives you no information about whether or not H_2 will occur. The same can be said about the second roll of a pair of dice, say, where rolling a 7 (or any other value) does not appear to affect what will happen when we roll again. In these cases, we describe our feelings by saying that the second outcome is *independent* of the first outcome. We can include this notion of independence in our mathematical model in the following way: since the conditional probability of H_2 given

H_1 is

$$P(H_2/H_1) = \frac{P(H_2 \cap H_1)}{P(H_1)},$$

the intuitive notion of independence suggests that if the occurrence of H_1 has no effect on the occurrence of H_2, this conditional probability should be the same as the unconditional probability. In symbols,

$$P(H_2/H_1) = P(H_2). \tag{5.1}$$

Put formula 5.1 into the left-hand side of the conditional probability formula above and multiply both sides of the resulting equation by $P(H_1)$ to obtain the famous *product rule* for independent events

$$P(H_2) \cdot P(H_1) = P(H_2 \cap H_1).$$

Events that do not satisfy the product rule, or equivalently, formula 5.1, are called *dependent*. This notion of independence introduced into our mathematical model turns out to be very fruitful. Most of the classical theory of probability was done under assumptions of independence; it is only relatively recently in the subject that various forms of dependence conditions have been studied extensively.

Let's note an interesting symmetry arising from the mathematics. We have said that H_2 is independent of H_1 because H_1 happened first and intuition demands that a first event may or may not affect the occurrence of a second event, not the other way around. Nothing, however, prevents us from considering $P(H_1/H_2)$, the conditional probability of a head on toss 1 given that a head on toss 2 occurred. Evaluate this using the conditional probability formula while at the same time assuming H_2 is independent of H_1 to get

$$P(H_1/H_2) = \frac{P(H_1 \cap H_2)}{P(H_2)} = \frac{P(H_1) \cdot P(H_2)}{P(H_2)} = P(H_1),$$

which is to say that our assumption of H_2 independent of H_1 implies that H_1 is independent of H_2, so the idea of independence is symmetric: as soon as a first event is known to be independent of a second, the second is automatically independent of the first. Of course, independence or dependence in our model simply means that a conditional probability is equal to an unconditional one or is not, and is not required to make intuitive sense in the real-life application of the model: what does it mean, you may ask, for a first toss to be influenced or not by a second toss? As mathematicians, we really don't have to worry about this question. As philosophers or physicists, we may find this interesting to speculate upon. The mathematics does not distinguish between the forward and backward directions of time. Because of this symmetry, we can simply say of two events that they are independent, without having to specify which of the two events is the conditioning one.

Now suppose the coin is tossed a third time with H_3 the event "head on third toss." For any pair chosen from the three events H_1, H_2, H_3, their independence implies the product rule for their probabilities. But the notion of independence for the three events requires something more: we want to express the idea that the probability of H_3 is unaffected not just by the knowledge of the occurrence of H_1 and H_2 separately but also by the occurrence of $H_1 \cap H_2$. To see where this idea leads, consider the following formula

$$P(H_1 \cap H_2 \cap H_3) = P(H_1) \cdot P(H_2/H_1) \cdot P(H_3/H_1 \cap H_2). \qquad (5.2)$$

Checking this formula is exercise 5 of Chapter 3, but we verify it now. The truth of formula 5.2 can be seen by observing first,

$$P(H_3/H_1 \cap H_2) = \frac{P(H_1 \cap H_2 \cap H_3)}{P(H_1 \cap H_2)}$$

by the conditional probability formula, and then expressing $P(H_2/H_1)$ also by the conditional probability formula, and finally substituting into the right-hand side of formula 5.2 to obtain

$$P(H_1) \cdot \frac{P(H_1 \cap H_2)}{P(H_1)} \cdot \frac{P(H_1 \cap H_2 \cap H_3)}{P(H_1 \cap H_2)} = P(H_1 \cap H_2 \cap H_3),$$

which indeed gives the left-hand side of the relation. So we have checked the truth of formula 5.2.

If we assume the three events in formula 5.2 satisfy the intuitive notion of independence whereby the probability of any outcome of a toss of the coin is unaffected by the knowledge of any of the other outcomes, then we can assert

$$P(H_2/H_1) = P(H_2) \qquad \text{and} \qquad P(H_3/H_1 \cap H_2) = P(H_3).$$

Put these expressions into the right-hand side of formula 5.2 to get the product rule for three events

$$P(H_1 \cap H_2 \cap H_3) = P(H_1) \cdot P(H_2) \cdot P(H_3).$$

After this introductory exploration into the idea of independence, we are ready for a precise definition for our mathematical model. Consider a sequence, possibly infinite, of events $A_1, A_2, A_3, \cdots$. We say that the events of the sequence are *mutually independent* (or just *independent*) if for any finite subset of these events the product rule holds, that is, if the probability of the intersection equals the product of the probabilities. So, for example, if A_3, A_8, and A_{41} are all defined, then it is necessary that

$$P(A_3 \cap A_8 \cap A_{41}) = P(A_3) \cdot P(A_8) \cdot P(A_{41}).$$

The above scheme could model, for example, successive tosses of a coin or successive rolls of a pair of dice, where the event A_i could be any event defined just in terms of the ith toss or roll, for instance, a head on toss i or snake eyes on roll i. The repetitive acts needed to generate the independent sequence of events are often called *independent trials*. The above mathematical definition is a formal way of stating that we are modelling a situation where no information is obtained about the outcome of any trial based on knowledge of the outcomes of any of the other trials.

We should, of course, realize that an actual case of repeated tosses of a coin may or may not be reasonably modeled by our abstract independent sequence. If, for example, I am able to control my tossing technique so that I can get what I want, the outcome of trial 2 may depend on what was obtained on trial 1. There has also been a philosophical argument claiming an actual sequence of coin tosses has a "memory," and if a lot of heads, say, occurs, the sequence tries to even things up by getting a tail more often. Such willful sequences would, of course, be dependent. There is no empirical justification for such a theory. In fact, experience seems to show just the opposite, since many gamblers have believed in sequences with memory and betted accordingly but still lost. On the other hand, the conclusions arrived at by the independence assumption are amply borne out by experience.

5.2   Waiting for the first head to show

Let us now consider a sequence of independent trials of tossing a coin. Moreover, suppose there is a number $p > 0$ such that at any trial the probability of a head is equal to p, and therefore the probability of a tail at any trial is equal to $q = 1 - p$. (If $p = .5$ we have the familiar case of a *fair* coin, where head and tail each have the same probability .5 of appearing.) For each positive integer i, consider the event

$$A_i = \{\text{head occurs for the first time at trial } i\}.$$

How do we calculate the probability of A_i? If a head occurs for the first time at trial i, then the $i - 1$ trials that preceded trial i each resulted in a tail. If $i > 1$, then A_i is the same as

$$T_1 \cap \cdots T_{i-1} \cap H_i,$$

where T and H stand for a tail and head occurring, repectively, at the designated trials. By the product rule for intersections of independent events, the above intersection has probability $q^{i-1}p$. If $i = 1$, this relation is still true since $q^0 p = p = P(A_1)$. Fine—we have just used the product rule to find the probability of an interesting type of event, where we are concerned with something happening for the first time. This type of event is important because it often allows us to decompose more complicated events into disjoint unions of such events. As an example, consider the event

$$H = \{\text{head occurs on at least one trial}\}$$

in the sequence, which we take to be infinite. H can be written as the infinite union of the events A_i:

$$H = A_1 \cup A_2 \cup \cdots, \tag{5.3}$$

because a head occurs at least once if and only if at least one of the events A_i occurs. But since the events A_i are also disjoint (you can't get a head for the first time at two different trials simultaneously), and since the probability of a union of disjoint events is the sum of their probabilities even when the union has an infinite number of terms as it does here, using the evaluation of $P(A_i)$ calculated above we now get

$$P(H) = p + qp + q^2p + \cdots. \tag{5.4}$$

The sum on the right-hand side is called an *infinite series*, and if you remember your high school algebra you recall that it is a very nice kind of infinite series called a *geometric series*. What makes it nice is that if $-1 < q < 1$ we can actually add up *all* the terms of the infinite series in formula 5.4 to get a number that we call the *sum* of the series. Well, what does it mean to add up an infinite number of terms, anyway? What we do is pick any finite positive integer n and consider the *finite* sum of the first n terms of formula 5.4 which we call S_n. This gives us no trouble because we are just dealing with the finite process of adding up a finite number of terms. We do this for each n and as n gets larger and larger (mathematicians say n tends to infinity) the sums S_n will (if $-1 < q < 1$) get closer and closer to a definite number, called the *limit* or *limiting value* of S_n. This value is *defined* to be the sum of the infinite series. This phenomenon of numbers getting closer and closer to some value as something else is changing (here it is the subscript n which is getting larger) is of fundamental importance in mathematics. Mathematicians use the term *convergence*—for example, the sequence S_n, the partial sums of the series, is said to converge to the sum of the infinite series, and then the series is called *summable*.

Of course, not all infinite series have the nice property of having a sum in this sense. The series

$$1 + 1 + 1 + \cdots$$

is not summable since the partial sums get large without bound, and so cannot settle down to get close to a limit. The geometric series of formula 5.4 is summable and according to our probability rules gives us the value of $P(H)$. How do we find the sum of a geometric series? Each geometric series starts with a first term (p above), and then each term is obtained from the preceding one by multiplying by a fixed quantity (q above). If the geometric series satisfies $-1 < q < 1$ [this holds true in formula 5.4], the sum is easy to get; it is always the first term divided by $1-q$ (the derivation of this is in any elementary algebra textbook). For formula 5.4 we get

$$\frac{p}{1-q} = \frac{p}{p} = 1.$$

Here the fact that $p = 1-q > 0$ is important; if $p = 0$, we would be dividing by 0 in the preceding relation, and this, as everyone should realize, is one of the supreme no-no's of mathematics.

What we have just proved is the following: suppose you keep tossing a coin in independent trials without stopping, and suppose the coin has a fixed probability $p > 0$ of getting a head at each trial. Then you will certainly (that is, with probability 1) obtain a head at least once in your sequence. (If $p = 0$, the series of formula 5.4 sums to 0 and we can never get a head, with certainty.) Another way to say this is that the complementary event "tail at all trials" has probability 0. Of course, this is a theoretical result; it is impossible in practice to toss a coin an infinite number of times. But such results can still give valuable practical information. Even though I can't toss a coin an infinite number of times, I can toss it a large number of times, say N. It follows that the probability that at least one head shows in N trials will be very close to 1 if N is large enough, and that as N gets larger, the probability gets even closer to 1. So this result about an event impossible in practice involving an infinite number of trials has useful things to say about events that can indeed happen.

5.3   On the likelihood of alien life

The tossing of a coin is an example of what the probabilist calls a sequence of *Bernoulli trials*, an experiment of independent trials with two possible outcomes at each trial, which we can call "success" and "failure" with respective probabilities p and $q = 1 - p$ (named after James Bernoulli of the distinguished Bernoulli family which produced several generations of outstanding mathematicians). If getting a head is identified with success and getting a tail with failure, then tossing a coin fits the model of Bernoulli trials. So do many other situations. For example, a machine that stamps out washers may produce washers in two possible states, good ones or defective ones. Then success may correspond to either of these states to fit the Bernoulli model. Similarly, a person exposed to a disease may or may not come down with the disease, rolling a pair of dice may or may not give a 7, a birth may yield either a male or a female; all of these instances illustrate the Bernoulli scheme. Now go back to Section 5.2 and note that the argument there proves: the probability of at least one success in an infinite sequence of Bernoulli trials with success probability $p > 0$ is 1 (just change the words "head" and "tail" to "success" and "failure" in the discussion).

Let us now ask: what is the probability of at least one success in a fixed number N of trials. In the preceding section, we saw that the probability of at least one success in an infinite number of trials is 1 (if $p > 0$) by adding up an infinite series. The present problem will be easier—we will only be required to add up the first N terms of the series of formula 5.4 to get $P(S_N)$. The event "at least one success in N trials" is given by the finite union

$$A_1 \cup A_2 \cup \cdots \cup A_N$$

rather than the infinite union given by formula 5.3, where now we interpret A_i to be "first *success* at trial i" rather than "first head at trial i." Thus the answer to the question is given by

$$p + pq + \cdots + pq^{N-1} = \frac{p - pq^N}{1 - q} = 1 - q^N,$$

the finite analog to formula 5.4. Here you have to know the formula from elementary algebra for summing a finite geometric series.

It is instructive to get this result by another, slicker method which does not require you to sum a series. Let us find the probability of no successes in N trials. This is the probability that each trial results in failure; the product rule gives this value to be q^N. The complement to the event of no successes in the N trials is that there is at least one success, and the rule on complementary events yields the result $1 - q^N$, as before. As N gets large, getting at least one success has a probability approaching unity, which simply means that the sum of the finite series above is very close to the sum of the infinite series of formula 5.4 if you add up a large number N of terms. This was also observed at the end of the last section.

According to many scientists, it is extremely unlikely that Earth is unique in the universe in supporting intelligent life. An argument for that viewpoint follows from what we have just done. There are certainly many stars in the universe (not an infinite number, however) with planets nearby where life would be conceivable. Consider each system, like our own solar system, to generate a Bernoulli trial: success if intelligent life exists, failure otherwise. Assume these Bernoulli trials are independent with the same probability p of success. The probability of intelligent life in a system, p, may be very small but it is positive. Now take a large number N of these solar systems, not including our own. What we have just calculated can be interpreted in the present case to assert that the probability of at least one of the N solar systems supporting intelligent life is $1 - q^N$, which is very close to 1 when N is large. Of course, this argument depends on your willingness to assume the applicability of the Bernoulli model of independent events to the set of solar systems.

5.4   The monkey at the typewriter

The story of the monkey at the typewriter is perhaps one of the most famous tales (or perhaps I should say "tails") of probability. It goes like this: a monkey is seated at a typewriter and randomly hits keys in an infinite sequence of independent trials. He produces, by this process, a neverending string of randomly selected characters from the keyboard. The story concludes with the assertion that the monkey will eventually type out the complete works of Shakespeare with certainty, that is, with probability equal to 1. In this section, I want to convince you of the truth and reasonableness of this proposition if you understand it in the right way. The statement is really an assertion about a probability model, not about a real monkey and typewriter. A real monkey will not keep pecking away indefinitely—he is a reasonable creature who will soon tire, toss the typewriter on the floor, and go off looking for a banana.

Perhaps the statement of the problem should be modernized by placing the monkey in front of a word processor; in any case, we won't worry about who puts more paper into the printer. Let's suppose there are M keys on the keyboard and the monkey can hit any one of these with equal probability M^{-1} at each trial. We must then decide on a certain (long) sequence of characters, which we shall refer to as the works of Shakespeare. In other words, the works of Shakespeare for us is a string of characters which, when read through, gives us all the plays and sonnets in some particular order. Now the number of characters in the works of Shakespeare is a finite number T, say. Consider the monkey typing away for T trials. For the monkey to type out the works of Shakespeare, he must type at each trial the unique character on the keyboard belonging in that position. But the probability of typing the "correct" character at any trial is M^{-1}. By independence, the probability of typing the "correct" character at each of the T trials is M^{-T} (all the correct characters typed in correspond to an intersection of events whose probability is the product of the individual probabilities). The most important thing to notice at this point is that M^{-T}, while very, very small, is nevertheless a positive number; call it a.

Now let us watch the monkey pecking away at the keyboard, and I will choose a trial to call the first trial and observe what the monkey produces in the first T trials. At the Tth trial I will be able to say whether or not the works of Shakespeare have been produced (you will, of course, know the first T trials must result in a *failure* to produce Shakespeare's works as soon as the monkey types his first incorrect character). The works have been produced only if each character typed matches the correct character for that position; if a single character is wrong, the works of Shakespeare have failed to be produced in spite of the fact that the monkey may have written out *Hamlet*, say, perfectly. Let's say that we have a success in the first T trials if the works of Shakespeare have been produced and a failure if they have not been. Then the probability of success in the first T trials is $a > 0$, and the probability of failure is $1 - a$.

The monkey, of course, keeps on typing, so I can start observing again what the monkey produces from trial $T + 1$ to trial $2T$; this is the second segment of T trials that comes right after the first segment from trial 1 to trial T. This second segment of T trials consists of independent trials with the same probability M^{-1} of hitting the correct key at each trial, so clearly the probability of success on this segment is also a (that is, the works of Shakespeare are produced on the segment from trial $T + 1$ to $2T$ with probability a).

Continue in this vein, next considering the segment from trial $2T + 1$ to $3T$; it again has probability of success a. So the monkey is involved in an infinite sequence of trials and we are breaking them up into non-overlapping segments where each segment has T trials and where we observe for each segment whether or not success has occurred. It is easy to see that the general ith segment stretches from trial $(i - 1)T + 1$ to trial iT. Define the events

$$S_i = \{\text{success occurs on the } i\text{th segment}\}.$$

What we claim is that the events S_i are independent. In fact, the ith segment involves trials that don't overlap with any trials from the jth segment for $i \neq j$. Since the individual trials form an independent sequence, it is intuitively reasonable that these disjoint segments of trials should be independent: whatever happens on one segment gives you no information about what happens on another. (A mathematical proof of this intuitive fact would require a quantitative argument showing the truth of the product rules at the basis of the definition of independence.) Once the independence of the S_i is accepted, we are over the hump and the problem starts to resemble things we did just a little while ago. Each segment can be considered as a single Bernoulli trial where success on segment i is the event S_i, and success has positive probability a, so the result on waiting times we found in Section 5.2 can be applied to get an immediate proof of the certainty of the monkey's ultimate triumph. To spell out matters in a little more detail, the probability of the *first* success occurring on segment i is $(1 - a)^{i-1} \cdot a$ because the first $i - 1$ segments resulted in failure, and the S_i are independent. Let S be the event that success will occur on at least one segment, that is, the works of Shakespeare are produced on some segment. S is the disjoint union of the events defined by the first success on segment i [this is similar to formula 5.3] and $P(S)$ is

$$a + (1 - a)a + (1 - a)^2 a + \cdots$$

[this is similar to formula 5.4]. Again, it is easily seen that this geometric series sums to 1 (remember $a > 0$), which is what we have been trying to show, namely, the certainty of the monkey ultimately typing the works of Shakespeare.

The solution can be arrived at slightly differently. If a failure occurred on each of N segments, a situation similar to the one in Section 5.3 would prevail. N failures has probability $(1 - a)^N$. For N growing larger, this probability shrinks to zero, so the probability of the event that there is at least one success is very large in N trials; that is to say, as the number of segments increases, the probability that at least one represents a success (the works of Shakespeare) converges to 1. This is equivalent to saying that $P(S) = 1$.

If you find it hard to believe that the monkey will eventually type out the works of Shakespeare with certainty, let me dismay you some more: a theorem can be proved asserting that the monkey will type out the works of Shakespeare not only once but actually *infinitely often* with certainty. You may feel better, however, when you realize how long it is going to take, on the average, for Shakespeare's works to be written out just once. Most likely longer than the sun will survive as a living star, so the monkey and equipment will have to be moved elsewhere.

5.5   Rare events do occur

The story of the monkey typing Shakespeare's works may be amusing, but it teaches an important lesson: rare events do indeed occur. A rare event is one with small probability. The same argument used for showing the certainty of the monkey's success at writing all of Shakespeare also shows that any rare event will eventually occur with certainty if the experiment producing it is repeated independently forever. How long do we have to wait? As we shall see in Chapter 7, if success has probability p, then the average waiting time until the first success in a sequence of Bernoulli trials is about p^{-1} trials. This gives us a way to estimate the waiting time for the monkey to succeed, and as we have noted, this is not an event to be expected during the lifetime of the solar system. But there are rare events that are not so extreme and probability models more applicable to real life.

For example, suppose a craps shooter rolls 7 ten times in succession. This would be an amazing run of luck; we would most likely think in terms of something wrong with the randomness of the game, perhaps some cheating. After all, a run of ten successive 7's has a probability roughly equal to $l = 1.6 \cdot 10^{-8}$, an extremely small number (to write 10^{-8} in standard notation, write 1.0 and move the decimal point eight places to the left). The theory, however, tells us that, in a perfectly legitimate random model, runs such as this should occasionally crop up. The average waiting time, in fact, will be about 625 million trials (the reciprocal of l times a factor of 10 due to the number of trials in a run). From this point of view, the ten successive 7's may seem a little less startling if viewed as part of the action of all the busy craps tables in the world. If we write down the results of all the rolls at all of those tables for a year preceding the occurrence of the ten 7's, we may interpret the run of 7's as part of a much longer sequence in which we have waited a reasonable time to observe the phenomenon.

In general, if a rare event is observed to occur, it may indicate a breakdown of the model or it may simply be the effect of random fluctuation. Further study is necessary to decide between these possibilities. Here is an example in the field of public health. The frequency of breast cancer in the female population gives an estimate of the distribution of the number of cases of the disease occurring in communities of various sizes. Suppose a cluster of cases is observed in a certain community, that is, the number of cases of the disease in the community is larger than one would expect from the estimated distribution. The cluster would constitute a rare event in terms of this estimated distribution. Is this cluster due to some environmental or other factors in the community or is it only a random fluctuation? Questions like this arise all the time and are often quite difficult to answer. An investigation must be conducted to determine whether any factors can be found making this community more dangerous than the average ones from which the distribution data were obtained. If this can be done, then the standard model and its distribution did not apply to this community, and that was the reason for the unusual observations. On the other hand, it may turn out that the standard model is indeed applicable and all we observed was a rare event: a cluster of cases in a community no more dangerous than the typical one.

5.6   Rare versus extraordinary events

Suppose a fair coin is tossed 100 times and a random assortment of heads and tails occurs which does not impress us as particularly interesting. Let us call this sequence 1. Now suppose we once again toss the coin 100 times and get a head on each of the 100 trials. Call this sequence 2. We are likely to be astonished at sequence 2. Yet there seems to be a bit of a paradox here, because both sequence 1, which we found uninteresting, and sequence 2, which we found startling, both have the same probability, 2^{-100}, of occurring. How can our surprise be explained?

Having small probability does not in itself make an event noteworthy if in fact it does occur. When you toss a coin 100 times, you must get *some* sequence, and whichever one you do get will be a rare event with probability 2^{-100}. Sequence 2 has something else about it, however, distinguishing it from sequence 1. When we toss a fair coin in independent trials, we expect to get heads roughly half the time and tails roughly half the time. Sequence 2 deviates from this expected result in so extreme a way that we find it hard to reconcile such a sequence with the outcomes resulting from tosses of a fair coin. In fact, as we shall see in Chapter 15 when we talk about statistical inference, if we observe sequence 2 we would strongly suspect the coin giving these outcomes was not really fair as proclaimed. Now suppose sequence 1 resulted in 55 heads and 45 tails in a random way without apparent pattern and is not very interesting. Suppose sequence 2 has the same number of heads and tails as sequence 1 except that *the first 60 tosses gave heads and the last 40 gave tails*—that would cause us to gasp a little. So even though sequence 2 has a reasonable number of heads and tails this time, the *pattern* in which they occur does not have the randomness associated with tossing a coin, and that would make sequence 2 extraordinary.

So we see that rare events, by themselves, are no cause for surprise. It depends on the entire context in which they are considered. The context causing surprise may be artificial, as the following example shows. Suppose we go back to sequence 1 as an ordinary looking sequence of 100 tosses of a fair coin. There is nothing startling in the occurrence of sequence 1, but what if, before we tossed, we had been told that if sequence 1 should occur we would be paid 2^{100}, whereas if anything else should occur we must pay $5. Now suppose sequence 1 *does* occur; we would be in a state of joyous shock. Sequence 1 has become surprising because we have focused on it and compared it to a much more likely competitor, the set of all other possibilities. So even though sequence 1 is no more unlikely than any typical sequence of outcomes of 100 tosses of a coin, its occurrence has become special and extraordinary purely through artificial means.

5.7   Exercises for Chapter 5

1. Chloe has two coins. Coin A is fair, with probability of a head $=1/2$, but coin B is biased, with probability of a head $=1/3$. Chloe tosses coin A and then in an independent trial tosses coin B, noting which side comes up in each toss. Describe the sample space and give the probability of each outcome. Find the probability of the events "at least one head" and "at least one tail."

2. A machine turns out troll dolls in Bernoulli trials where, on the average, 1 out of 1000 trolls produced is defective. Assuming the machine runs continuously forever, find (a) the probability that the first defective troll occurs after the 100th trial, (b) the probability that the first troll is defective and all future trolls produced are good, (c) the probability that a run of a million trials produces at least one defective troll.

3. Consider the following statement: "If you keep playing the Lotto game under the same conditions you must eventually win because you have a positive probability of winning at each play. So it makes sense to keep playing." Discuss the validity of the statement.

4. (*Return of the car and goats yet again.*) You play the car-goat game two times in independent sessions. The first time you don't switch

and the second time you do switch. What is the probability that you will win two goats? Two cars?

5. When Ringo drives to work he passes three traffic lights in succession. He has noticed that about 1/4 of the time each light has been green as he approaches the corner where the light hangs, about 1/4 of the time the first has been green and the other two red, about 1/4 of the time the second has been green and the other two red, and about 1/4 of the time the third has been green and the other two red. (a) Describe the probability space obtained by considering the possible colors of each of the three lights as Ringo approaches. (b) Let F, S, and T be the events "the first light is red," "the second light is red," and "the third light is red." Find the probabilities of F, S, and T and of the events $F \cap S$, $F \cap T$, $S \cap T$, and $F \cap S \cap T$. Conclude from this that the events F, S, and T are not independent even though any pair of these three events are independent.

6

A Little Bit About Games

In gambling, one thing you should never do is take something
for granted.
John Scarne, *Scarne's Guide to Casino Gambling*

6.1 The problem of points

In this chapter, we discuss several games; more along this line will follow
in the next chapter. The first problem, solved by both Pascal and Fermat,
goes back to the earliest days of probability as a formal theory. Suppose two
people are playing a game with the winner receiving prize money at the end.
If the game is forced to end before either player wins, how should the prize
money be divided between the players? Pascal introduced the principle that
the prize money should be divided in proportion to each player's conditional
probability of winning if the game were to be continued, given the score
when the game is forced to end. Suppose, for example, that the plays of the
game constitute a sequence of Bernoulli trials where A wins a point with
probability p (success) and B wins a point with probability $1 - p$ (failure),
and n points are needed to win. We will not derive the general formula
but will give the solution for the case where A has $n - 1$ points and B has
$n - 2$ points. Then A needs one point to win and B needs two points. A
can win in two ways if the game were to be continued at this moment: (1)
A can win the next point, and (2) B can win the next point and A can win
the succeeding point. This gives the value $p + p(1 - p)$ for the conditional
probability of A winning. If $p = 1 - p = .5$ and the purse is \$100, then,

according to Pascal's principle, A should receive \$75 and B \$25.

6.2 Craps

Craps is played with a pair of dice. The player (sometimes called the "shooter") rolls once. If the dice show 7 or 11, she wins. If the dice show 2, 3, or 12, she loses. If the dice show any other value, this number is known as the gambler's "point." She must now keep rolling the dice until either she gets 7 before her point appears, in which case she loses, or else gets her point before 7 appears, in which case she wins. In addition to the shooter, most real-life craps games have a host of other people betting on the shooter's game.

We are going to calculate the probability of the event "the gambler wins at craps." This is an interesting game to analyze because the sample space is rather complicated. A typical element of the sample space can be considered an n-tuple $(x_1, x_2, \cdots, x_n)$ of n rolls of the dice, where the entry x_i denotes the number appearing on roll i, and the game ends at roll n. Using this notation, the simplest elements of the sample space can be written: (7), (11), (2), (3), (12). Suppose the first roll is 4; this becomes the gambler's point. The sample space contains elements of the form $(4, x_2, \cdots, x_n)$ where the term x_n is either 7 or 4 and the terms x_2 through x_{n-1} must be different from both 7 and 4. The totality of such elements can be described as the event "the first roll is 4, and the game ends at roll n." Let us calculate the probability of the event "the first roll is 4, and the gambler wins at roll n." For this to happen, there must have been a sample point of the type $(4, x_2, \cdots, x_{n-1}, 4)$, where the final 4 is in the nth position, and the $n - 2$ terms between the two 4's may not be either 7 or 4. A roll of 4 on two dice can occur in three ways out of 36 possible rolls, so the probability of the initial 4 as well as the terminal 4 is 3/36. The probability of each of the $n - 2$ terms between the 4's is 27/36, since nine of the 36 rolls are excluded (six ways for 7, three ways for 4). By independence of the rolls, the product rule gives us

$$P(\text{the first roll is 4, and the gambler wins at roll } n) = \left(\frac{3}{36}\right)^2 \cdot \left(\frac{27}{36}\right)^{n-2},$$

valid for $n \geq 2$.

Now suppose we want to find the probability of the event "the first roll is 4, and the gambler wins at some time." This event is the disjoint union of the events that the gambler wins at time n for the times $n = 2, 3, \cdots$. We are already experts at this sort of thing [if you don't agree with this, go back and look at formulas 5.3 and 5.4 of Chapter 5], so we know

$$P(\text{the first roll is 4, and the gambler wins at some time}) \qquad (6.1)$$

$$= \left(\frac{3}{36}\right)^2 \left(1 + \left(\frac{27}{36}\right) + \left(\frac{27}{36}\right)^2 + \cdots\right) = \frac{1}{36}$$

by summing the geometric series. The above reasoning shows the pattern for solving the original problem: calculate the probability for winning for each of the possible points in a manner similar to what was just done. Add these values together with the probability of winning by making an initial 7 or 11 and you have the overall probability of winning. We now take each of the points in turn and find the analog of formula 6.1. There are three ways of rolling 10 just as there are for 4, so the numbers on the right-hand side of formula 6.1 are the same and the sum is again 1/36. There are four ways of rolling either 5 or 9, so each of these points gives the general term in the analog of formula 6.1 to be

$$\left(\frac{4}{36}\right)^2 \cdot \left(\frac{26}{36}\right)^{n-2}$$

and the sum of this series is 2/45. Finally, there are five ways of rolling each of the points 6 and 8, so the general term of the series analogous to the right-hand side of formula 6.1 has general term

$$\left(\frac{5}{36}\right)^2 \cdot \left(\frac{25}{36}\right)^{n-2}$$

and the sum is 25/396. The probability of winning by getting either 7 or 11 on the first roll is 8/36 (six ways for 7, two ways for 11). The probability of winning at craps is therefore

$$\frac{8}{36} + 2 \cdot \left(\frac{1}{36}\right) + 2 \cdot \left(\frac{2}{45}\right) + 2 \cdot \left(\frac{25}{396}\right) = .492927\cdots.$$

6.3 Roulette

Roulette is perhaps the most glamorous and romantic of casino games; we have all seen the spinning wheels and intent players in the movies. Here is a rough description of the American version of the game. There is a wheel divided into 38 slotted sectors of equal size. Thirty-six of these are marked by the numbers 1 through 36, with 18 numbers colored red and the other 18 black. The remaining two sectors are green in color and marked 0 and 00. The croupier spins the wheel and then spins the ball in a groove on the wheel. Eventually the ball slows down and falls off the groove into one of the 38 slotted sectors. Various bets can be made about the number on which the ball lands, for example, red or black, odd or even, any particular number, or groups of numbers such as "low" numbers (1 to 18) or "high" numbers (19 to 36). The mathematics of roulette is very simple. Here are a few calculations.

$$P(\text{red}) = P(\text{black}) = P(\text{odd}) = P(\text{even}) = 18/38 \approx .474,$$
$$P(\text{any number}) = 1/38 \approx .026.$$

6.4   What are the odds?

Gamblers don't usually talk in terms of probabilities of events. They prefer to talk in the language of *odds*, the ratio of the number of unfavorable ways to the number of favorable ways. In craps, the number 7 can be rolled in six out of 36 ways, so there are 30 unfavorable ways that 7 cannot come up. The odds are therefore 5 to 1 against 7 appearing in one roll of a pair of dice. The odds against snake eyes are 35 to 1. In roulette, the odds against any particular number appearing are 37 to 1. If the odds against an event are i to 1, it means that a fair payoff if you win the bet would be $\$i$ for every \$1 you bet.

Payoffs in a casino are not in accordance with fair odds, of course; they are always somewhat less. The odds against a shooter winning at craps are about 1.028 to 1 (the shooter wins about 493 times out of 1000 when we interpret the probability of winning given in Section 6.2 as a relative frequency). But the payoff is only at even money; that is, the house gives you only \$1 for every \$1 you bet. In roulette, if you bet \$1 on a particular number and win, the house only gives you \$35 rather than the \$37 they should pay you according to the true odds. In other words, they are paying you as though only 36 numbers were on the wheel. We can say that in roulette 35 to 1 are the *payoff* odds in contrast to the true odds.

We will speak more about games in the next chapter after the important ideas of random variable and expectation are developed. Fair odds would give you the payoff in what we call a fair game. All casino games are inherently unfair, biased against the gambler. This gives the casino its *edge*, the way it makes money in the long run because of the laws of probability. For example, the two green sectors in roulette give the house its edge. Sometimes gamblers have sought to eliminate the casino's edge by trying to obtain additional information about the game and therefore increase their conditional probability of winning—see Section 7.6.

6.5   Exercises for Chapter 6

1. Suppose you play a single game of craps each day of your four-day vacation. Find the approximate probability that you will win at least one of the four games.

2. Suppose the plays of a game are a sequence of Bernoulli trials where my opponent and I have probabilities 1/3 and 2/3, respectively, of scoring a point, and the first one to score 21 points wins. There is a

$100 purse for the winner. If the game must end prematurely when I have 18 points and my opponent has 20 points, use Pascal's principle to calculate how the purse should be divided.

3. What are the odds against rolling snake eyes in one roll of the dice? What are the odds against tossing at least one head in three tosses of a fair coin?

4. There are four red balls and one black ball in an urn. You choose a ball at random, replace it, choose a second ball at random, and so on until you get for the first time a ball whose color is the same as the color of the first ball selected. Find the probability that the game will end when you select a black ball. Find the probability that the game will end when you select a black ball, and that it will take more than three selections from the urn to accomplish this.

5. Anna is playing roulette and wins with the ball on the red 7. What is the probability that only red numbers appear before the red 7 wins again?

6. Your friend the gambler has three cards. Each card has a mark on each side. One card has a red mark on each side, one card has a black mark on each side, and one card has a red mark on one side and a black mark on the other. One of the cards and one of its sides is chosen at random so that the mark on only that one side is visible. This mark is seen to be red. You know the other side of the chosen card can be black or red, and the gambler says each of these possibilities is equally likely. He wants to bet even money that the other side is red. Is this a reasonable bet for you to make?

7
Random Variables, Expectations, and More About Games

> When haughty expectations prostrate lie,
> And grandeur crouches like a guilty thing
> William Wordsworth, from *Sonnet 21*

7.1 Random variables

Suppose we toss a coin three times in independent trials, with probability p of getting a head at each trial. On the average, how many heads can we expect to get? Right now this question does not have a precise meaning for us; what do we mean by "average" or "expect"? Most likely we do have a rough idea of the meaning of the question. Three tosses of a coin will result in anywhere from 0 to 3 heads, so the answer must be some number in that interval. The first step in making the question precise is the definition of the term *random variable*. A random variable is a correspondence that assigns to each outcome in a sample space a unique number. (Mathematicians more generally refer to such objects as *functions*.) For example, in the above set-up of three tosses of a coin, let X = the total number of heads obtained. The table below shows the assignment of a value for X to each of the eight possible outcomes.

$$(H, H, H) \longrightarrow 3 \qquad (H, H, T) \longrightarrow 2$$
$$(H, T, H) \longrightarrow 2 \qquad (H, T, T) \longrightarrow 1$$
$$(T, H, H) \longrightarrow 2 \qquad (T, H, T) \longrightarrow 1$$

$$(T, T, H) \longrightarrow 1 \qquad (T, T, T) \longrightarrow 0$$

The number to the right of any arrow is the number of heads appearing in the triple to the left of it; it is the value of X assigned to the given triple.

A random variable has a *distribution*. This is the list of each possible value of X with the probability of attaining that value. The distribution can be computed from the above table easily using the given set-up of independent tosses with probability p of heads. For instance, each triple with exactly two heads has probability $p^2(1 - p)$, and there are three such triples, so the probability of $X = 2$ is $3p^2(1 - p)$, and so on for each possibility. The distribution of X can be written

$$P(X = 0) = (1 - p)^3, \qquad P(X = 1) = 3p(1 - p)^2,$$
$$P(X = 2) = 3p^2(1 - p), \qquad P(X = 3) = p^3.$$

Here is another example of a random variable. Let's say I am the shooter at a casino game of craps where the house is betting even money that I lose; this means that if I bet $\$n$ I pay this to the house if I lose and if I win the house pays me $\$n$ (see Chapter 6). To make things simple, suppose the amount of the bet is $\$1$. Let $X = $ my winnings after one game. X has possible values 1 and -1 (a loss is expressed as a negative win). Let us take the probability of winning at craps to be .493. The distribution of X is then $P(X = 1) = .493, P(X = -1) = .507.$

The above two examples exhibit *discrete* random variables, that is, random variables whose associated probability space is discrete. Such random variables have either a finite number of possible values or an infinite number that can be counted off using the positive integers. An example of a discrete random variable with an infinite number of values is given by letting $X = $ first time success occurs in a sequence of Bernoulli trials. The event $X = i$ means that the first $i - 1$ trials resulted in failure and the ith trial gave success. The probability of this event, which is precisely the value of $P(X = i)$, has already been calculated in Section 5.2 (where success is equivalent to obtaining a head) and has the value $q^{i-1}p$ (here, as usual, p is the probability of success and $q = 1 - p$ is the probability of failure—from now on when we discuss Bernoulli trials, p and q will always denote these quantities). The distribution of X is called a *geometric* distribution because the probabilities $q^{i-1}p$ are terms of a geometric series.

Any statement about the values of X can be traced back to describe an event in the sample space. In the first example above, $X > 0$ corresponds to the event "all triples containing at least one H." This event has the same probability as $X > 0$; it can be calculated from the probability space directly or from the distribution of X: $P(X > 0) = 1 - P(X = 0)$. It should also be evident that if you have a random variable and you add all the probabilities up for all possible values, you must get 1. This must be the case since each outcome in the sample space has some value of the random variable corresponding to it. Adding up the probabilities of all the values is the same as adding up the probabilities of all the outcomes.

7.2   The binomial random variable

The first random variable we considered involved tossing a coin three times and defining the variable to be the total number of heads obtained. A general version of this variable defines X to be the total number of successes in n Bernoulli trials. Each trial gives one of two possibilities, and a typical outcome is an n-tuple $(x_1, x_2, \cdots, x_n)$, where each x_i is either S or F, for success or failure, respectively. How many such outcomes are there? By the counting principle, you must multiply 2 by itself n times to obtain 2^n outcomes. Let's try to calculate $P(X = i)$, the probability of exactly i successes in the n trials. Any outcome with exactly i successes and therefore $n - i$ failures must have probability $p^i q^{n-i}$ by the product rule of independence. So if we know how many outcomes there are with i successes, the desired probability can be calculated by multiplication. The number of such outcomes is just the number of different ways we can write down an n-tuple with S appearing exactly i times and F appearing $n - i$ times. The calculation is similar to ones done in Chapter 2. Let's suppose we have n symbols we want to enter in some order in the n-tuple. There are n ways of selecting the first entry of the n-tuple, $n - 1$ ways of selecting the second entry, and so on, so the totality of ways of entering the symbols when they are all considered *distinct* is the product of all the integers between 1 and n inclusive, a value mathematicians designate by $n!$ (pronounced "n factorial") . But the symbols in this case are not all distinct; none of the S's can be distinguished from one another, nor can the F's. So $n!$ is too big and we have to divide through by a factor giving the number of ways we can order the S's and F's among themselves. The S's can be permuted in $i!$ ways (think of the S's as i distinct symbols; they can be written in the i spaces in $i!$ ways but all of these ways correspond to a single distinguishable pattern in our counting) and the F's in $(n - i)!$ ways. This yields the formula

$$\text{total number of } n\text{-tuples with } i \text{ } S\text{'s and } n - i \text{ } F\text{'s} = \frac{n!}{i!(n - i)!}.$$

The right-hand side of this formula is sometimes expressed as $C_{n,i}$, read "n choose i." We recognize it as the number of ways of choosing a set of i objects from a group of n. That this formula should appear in our calculation is natural since each distinguishable pattern of S's and F's obtained is determined by choosing from the n spaces of the n-tuple exactly i of them to write in the letter S. The distribution of X is therefore given by

$$P(X = i) = \frac{n!}{i!(n - i)!} \cdot p^i q^{n-i}.$$

This is called the *binomial distribution* and depends upon two parameters (that is, variables whose values define the distribution), the number n of trials and the probability p of success. If $n = 3$, the formula reduces to the distribution of the total number of heads obtained in three trials calculated directly in Section 7.1.

7.3 The game of chuck-a-luck and de Méré's problem of dice

The game of chuck-a-luck is played by rolling three dice. The gambler bets on one of the numbers 1 through 6. This number may appear zero, one, two, or three times; if it appears i times, the gambler receives \$$i$. The dice are assumed to act independently, so we think of each die determining a Bernoulli trial with success corresponding to the gambler's number turning up. The probability of success is $1/6$, and if we define the random variable X = amount paid by i appearances of the winning number, then X has a binomial distribution with $n = 3$, $p = 1/6$. The distribution of X is given by

$$P(X = i) = C_{3,i} \left(\frac{1}{6}\right)^i \left(\frac{5}{6}\right)^{3-i}.$$

We turn now to a problem that spurred the development of the theory. The Chevalier de Méré deserves a place in the history of probability perhaps not so much for solving any problem as much as for asking good questions. He was the one who called Pascal's attention in 1654 to the problem of the points discussed in Chapter 6. He also asked Pascal's advice about a problem of dice. It seems in those days there was a popular game in which the house would bet even money that a gambler throws at least one 6 in four rolls of a single die. Each roll is a Bernoulli trial with success equivalent to rolling a 6, and the probability of success is $1/6$. From the binomial distribution

$$P(\text{no 6's in four throws}) = \left(\frac{5}{6}\right)^4,$$

and so

$$P(\text{at least one 6 in four throws}) = 1 - \left(\frac{5}{6}\right)^4 = .517\cdots,$$

giving the house an edge. An old gambler's rule seemed to indicate that since the game of at least one 6 in four throws was favorable to the house the probability of at least one *double* 6 showing on 24 rolls of a *pair of dice* should still be favorable to the house. The idea was that the four throws of one die had to translate into $6 \cdot 4 = 24$ throws of two dice because two dice can come up in six times as many ways as one die. Yet de Méré did not believe the game with the pair of dice was favorable to the house. Some feel that his suspicions were aroused by gambling losses, others that he came to his conclusions through reasoning. There are arguments against both explanations. On one hand, he would have had to gamble a lot to get enough data to distinguish between extremely close probabilities. On the other hand, not much was known about probability calculations in 1654. So there is a little bit of a mystery here as to how de Méré noticed something puzzling in the first place, but he did, and he asked Pascal, who solved the problem. For us now, the problem is easy. Let a single roll of the dice constitute a Bernoulli trial with success corresponding to a double 6 appearing. The probability of success is 1/36, and from the binomial distribution

$$P(\text{no double 6's in 24 trials}) = \left(\frac{35}{36}\right)^{24}.$$

The right-hand side can be easily found by taking $24 \cdot \log(35/36)$, where "log" is the base 10 logarithm (use a calculator), and then taking the antilogarithm of this result (raising 10 to this power). We get, to three decimal places, .509, which means that the probability of at least one double 6 is approximately .491, confirming de Méré's suspicion that the game with two dice is not favorable to the house.

7.4   The expectation of a random variable

Now we are ready to introduce the all-important idea of the *expectation* or *expected value* of a random variable. For the moment, we restrict the discussion to discrete random variables. The basic idea is that, since a random variable has in general very many values, it would be most pleasant if there were some number giving the average of these values in some sense. The ordinary arithmetic average of a bunch of numbers is obtained by adding them all up and then dividing by how many numbers you have. This average for a random variable is not such a good idea because for a random variable it is the *distribution* that is important (that is, the probability information about the values), not the raw values themselves. For example, consider a random variable X with the two possible values 100 and 0 with probabilities .99 and .01 respectively. The arithmetic average is 50, but the probabilities indicate a probable value of 100. By changing the distribution so that 100 and 0 have probabilities .01 and .99, respectively, the probable value is now 0. What we want is a way to get an average probable value. We do this by taking a *weighted average, using the probabilities as weights*. This means that values with large probabilities count more and values with small probabilities count less in the average. To be precise, suppose a discrete random variable X has a distribution given by a list $P(X = a_i) = p_i$, namely, for each possible value a_i of X its probability is p_i. The *expectation* of X, written EX, is defined by

$$EX = a_1 \cdot p_1 + a_2 \cdot p_2 + \cdots,$$

the series having a finite or infinite number of terms depending on whether X has a finite or infinite number of possible values (in the infinite case we may run into problems because the series may not be summable. In this case, the expectation may not exist, but for the moment we won't worry about this). In words, to calculate the expectation of a discrete random variable, multiply each possible value by the probability of that value, and then add up all the terms you get. The expectation of X is sometimes called its *mean*. This use of the term *mean* must be distinguished from statistical usage when considering the arithmetic average of observations. The arithmetic average of observations, or data, is the *sample mean* (see the end of Chapter 12 and Chapter 15); what we are discussing now is often distinguished from the sample mean by calling it the *population* mean.

Armed with this definition of expectation, let's return to the question at the beginning of this chapter and answer it by calculating the expectation of the random variable X of the first example of Section 7.1. Using the distribution of X, the definition of expectation gives us

$$EX = 0 \cdot (1-p)^3 + 1 \cdot 3p(1-p)^2 + 2 \cdot 3p^2(1-p) + 3p^3 = 3p,$$

and if the coin is fair, $EX = (3)(.5) = 1.5$. So for a fair coin we expect 1.5, half of the 3 tosses, to be heads. Note that the expected value of a random variable need not be a possible value. You cannot get 1.5 heads; it is merely a number defined in a certain way. Similarly, if X is given as in the second example of winnings at craps, clearly

$$EX = 1 \cdot (.493) + (-1) \cdot (.507) = -.014.$$

For this random variable, the expected value is negative and we expect to lose about 1.5 cents in each game.

For another example, take a look at the variable X with the geometric distribution. The quantity X measures the time until the first success in a sequence of Bernoulli trials. From the distribution given in the previous section

$$EX = 1 \cdot p + 2 \cdot qp + 3q^2p + \cdots + i \cdot q^{i-1}p + \cdots. \qquad (7.1)$$

This is an infinite series but is not a simple geometric series, and it is not even clear that the series is summable. But not to worry. The right-hand side of formula 7.1 does indeed define a summable series, and the sum can be evaluated quite easily by a method depending upon calculus. There is, however, a cute little trick by which we can evaluate the sum with hardly any work (if we are willing to forgo the rigorous justification of the argument, that is). Define X^* to be the time of the first success that appears *after the first trial* . Notice

$$X = \begin{cases} 1, & \text{if first trial is success,} \\ 1 + X^*, & \text{if first trial is failure.} \end{cases}$$

Now, it seems reasonable to suppose that the expectation of X should be the sum of the expectations under each of these possibilities multiplied by the probability of the possibility; that is,

$$EX = 1 \cdot p + (1 + EX^*) \cdot q.$$

But X^* is exactly the same kind of random variable X is and has the same distribution. It therefore has the same expectation. So we get

$$EX = p + q + qEX = 1 + qEX,$$

and solving for EX gives us $EX = (1 - q)^{-1} = p^{-1}$, an elegant result with intuitive appeal. The expected number of trials before the first success occurs is just the reciprocal of the probability of success. If this probability is, say, $1/1000$, the frequency interpretation of probability says that roughly there is 1 success in 1000, or that we can expect a first success after trial 1000. In general, if the probability of success is small, then you have to wait a long time on the average, but if this probability is large, then your average wait is short. If $p = .5$, you expect two trials until the first success.

In the above discussion, we noted that since the two random variables X and X^* have the same distribution, then their expectations (assuming they exist) must be equal. This important relation is true in general: random variables with the same distribution must have the same expectation when the expectation exists. This follows from the definition of expectation, a quantity that depends only on the distribution function.

If X is a random variable with a uniform distribution, then the expectation of X is just the ordinary arithmetic average. For example, suppose you roll a fair die, each face having probability $1/6$ of appearing. Let $X =$ number appearing on the roll. Then

$$EX = 1 \cdot \frac{1}{6} + 2 \cdot \frac{1}{6} + 3 \cdot \frac{1}{6} + 4 \cdot \frac{1}{6} + 5 \cdot \frac{1}{6} + 6 \cdot \frac{1}{6} = 3.5,$$

the ordinary average.

The expectation gives a measurement of the center of a distribution of a random variable X. There are other ways of measuring a central value. The *median* is another such measurement, which is sometimes defined as the smallest value m such that $P(X \le m) \ge .5$. To take a simple example, suppose X is the total number of heads in four Bernoulli trials with a fair coin. The possible values of X are ordered: 0, 1, 2, 3, 4, with corresponding probabilities $1/16$, $4/16$, $6/16$, $4/16$, $1/16$, as is easily calculated. Starting from the left of the ordered values at 0 and adding the probabilities of the values, the first time .5 is exceeded is at $X = 2$, so 2 satisfies $P(X \le 2) \ge .5$ and is the smallest such value and is the median. A computation shows the expectation of X in this case is also equal to 2. But in general the median is not equal to the expectation; just take the case of three tosses of the coin rather than four. The possible total number of heads is 0, 1, 2, 3 with probabilities 1/8, 3/8, 3/8, 1/8, and the median is 1 but the expectation is 1.5. Note that the definition of median requires it to be a possible value, whereas there is no such restriction on the expectation.

Why are there different notions of central value, and which one should we use? The answer is that there are different ones for different purposes and the one to be used depends upon the purpose. The importance of the expectation as a measurement is justified by its appearance in the Law of Large Numbers, as we shall see in the next chapter. It is also an intuitively appealing measure of the average of the possible values of a random variable. On the other hand, the median is useful when you are interested in a notion of central value that guarantees a position in the middle of the distribution: it is roughly as likely for other values to be smaller than the median as larger; that is, about half the probability lies to the left of the median and about half to the right. If we are dealing with observations, we can define the *sample median*, which would be the middle observation or the average of the two middle observations when the data is arranged in increasing order. The median, or sample median, often turns out to be a more natural indicator of the center of a distribution than the expectation, for example, in statistical studies where the distributions of variables are largely unknown. Moreover, the median is an indicator much less sensitive than the mean to "outlier" or extreme values, observations that are very different from most of the others. Suppose, for instance, there are ten values, each with the same probability .1. If nine of these values are between 0 and 1 but the tenth is larger than one, then the median is the same regardless of the value of the tenth value. The mean, however, depends very strongly on all values (take the tenth value equal to 10^6, say). It is good to have several indicators for the idea of a central value—results can be compared using each of them, and the theory becomes richer.

7.5   Fair and unfair games

We want now to consider random variables that represent an individual's winnings in a game of chance. Let X = your winnings, where a negative value of X is the sum of any losses and possible fees required to play the game. Although the language of gambling used here may seem frivolous, the model to be described has wide application; after all, the purchase of any kind of insurance is a "game" where we "win" when we (or our heirs) collect, and our "losses" are our premium payments. A game is called *fair* if $EX = 0$. There are two types of unfair games, *favorable* when $EX > 0$ and *unfavorable* when $EX < 0$.

As we have observed, casinos prosper because the games played there are always unfavorable to the gambler. The game of craps described above has an expectation of $-.014$ and is unfavorable. Betting $1 on the red numbers in American roulette gives

$$EX = 1 \cdot (.474) + (-1) \cdot (.526) = -.052,$$

so your expected loss here is about 5 cents. From the point of view of expectation, the craps game is better for the gambler than roulette. Now let's consider the Lotto game discussed in Chapter 2. Suppose the purse is $10,000,000. This can be expressed more succinctly in so-called scientific notation as 10^7, that is, 1 followed by 7 zeros. Recall that for $1 you can buy two game panels, each of which has probability $1/25,827,165 \approx 3.8 \cdot 10^{-8}$ of winning the purse. It follows that you lose your $1 with probability $25,827,163/25,827,165 \approx 9.9 \cdot 10^{-1}$ and win the purse with probability $7.6 \cdot 10^{-8}$ (we assume, for simplicity, that there are no ties; that is, you are the unique winner who doesn't have to share the purse). Therefore, the random variable X representing your winnings at Lotto has expectation

$$EX \approx (-1) \cdot (9.9 \cdot 10^{-1}) + 10^7 \cdot (7.6 \cdot 10^{-8}) = -.99 + .76 = -.23 \,,$$

so your expected loss in this game is 23 cents on the dollar. Finally, let's turn to chuck-a-luck. If X is the number of times the number bet on turns up, and the gambler bets $1, and if Y is the gambler's winnings, by putting in the probabilities given in Section 7.3 we have

$$EY = (-1) \cdot P(X = 0) + 1 \cdot P(X = 1) + 2 \cdot P(X = 2) + 3 \cdot P(X = 3) \approx -.079.$$

In terms of odds, a payoff at the true odds corresponds to a fair game. To take an example, the true odds against any particular number in roulette is 37 to 1. This means the payoff is $37 for every $1 you bet. The expectation of your winnings is therefore

$$37 \cdot \frac{1}{38} + (-1) \cdot \frac{37}{38} = 0.$$

Of course, the casino does not give a payoff at the true odds. The payoff odds for a number in roulette is $35 for every $1 you bet. So the expected winnings is obtained by putting in 35 for 37 in the above equation. This comes out to an expected loss for the gambler of about 5 cents.

The Petersburg game is a theoretical game that goes like this: you toss a fair coin until you get a head. If this occurs at trial i, you receive 2^i. Let X be the payoff of the game. Then

$$EX = 2 \cdot (2^{-1}) + 2^2 \cdot (2^{-2}) + \cdots + 2^i \cdot (2^{-i}) + \cdots$$

is the expected payoff from the game. Each term of this infinite series gives 1, so the series is not summable—the partial sum of the first n terms is n, and there is no convergence of the partial sums to a nice number. That means there is no expectation in the precise sense although we might say that the expectation is *infinite* as a shorthand way to describe the steady growth of the partial sums beyond all bounds. The question that intrigued early workers in the theory was this: what would be a fair entrance fee to play the Petersburg game? No matter how much money you propose to pay, the game cannot be made fair since your expected winnings will be infinite. On the other hand, would you be willing to pay $2^{10} = \$1,024$ to play? If so, tails would have to come up for the first nine consecutive tosses before you could just break even, and the probability of this is approximately .00195.

The Petersburg game is sometimes called the *Petersburg paradox* because it seemed so strange to these early probabilists that on one hand the game is technically favorable for the gambler no matter how large an entrance fee is paid, and yet few people would be crazy enough to bet even the relatively tame amount of $1,024. The paradox is cleared up when it is realized that the rules of the Petersburg game implicitly assume the house must have an infinite amount of capital to be able to take on the gambler. This, of course, is impossible; the house only has a finite amount of capital it can lose before it goes broke. A realistic version of the Petersburg game would therefore have to end at a finite time (when the house reaches its maximum capacity to pay), and this would lead to the sum of a finite number of terms, namely, a finite number, for the expectation of the payoff. To get a fair game, the amount of this expected payoff is the entrance fee paid by the gambler. To get a game favorable to the house, any amount larger than this expected payoff would work.

How do the ideas about fair games apply to the problem of determining insurance premiums? Here is a simplified argument showing how a premium payment can be roughly determined using the notion of a fair game. Suppose a man 36 years old wants to buy $50,000 of term life insurance for a 20-year term. Over the years, much data have been collected on mortality for all age groups; these data have given rise to mortality tables used by insurance companies to estimate the probability p_i that an individual of age i will survive for 20 more years. The expected payoff by the insurance company on the insuree's life is $D = \$50,000 \cdot (1 - p_{36})$. Now, what is the expected income to the insurance company? Even if the insuree dies, he will have paid some premiums until his death. To simplify matters, we will neglect that income, and just consider the income to the company when the man lives the full term and pays all his premiums. If the yearly premium is a fixed amount $L/20$, the company collects a total amount L. The expected income to the company is therefore $L \cdot p_{36} - D$. If $L = D/p_{36}$, the game is fair, but since the company is in business to make money (like a casino) its game must be unfavorable to the insuree, and L will be chosen to make the expected income positive and give the insurance company its edge. We have here ignored, for example, other complications, like lapsed policies or the income the insurance company makes by investing its premiums.

An interesting question concerns the controversy sometimes arising about the way the probabilities p_i are determined from the data. Women in general live longer than men and so have different mortality statistics. If premium computations for a female are based on mortality tables from a female population, p_i should be larger than if the computations are based on tables from a general population including men, and so D and the annual premium should be smaller. This reflects the fact that the insurance company is taking less of a risk if it insures a woman rather than a man. Arguments like this have been used by different social groups as a plea for lower premiums; for example, non-smokers might ask to pay less for health insurance than smokers. On the other hand, similar reasoning could lead to high premiums or outright refusal to offer insurance to people in high risk groups, like cancer or AIDS patients. These are complicated non-mathematical questions depending on society's view of the role of insurance.

7.6   Gambling systems

Gamblers have learned the hard way over the centuries that there does not appear to be any way to beat the odds by implementing various systems. The games we have described depend entirely upon chance, and a negative expectation means you must lose in the long run (as we shall see in Chapter 8) if you play the game repeatedly under the same conditions. You may believe, however, that deciding when to play by stopping or skipping at certain times, or changing the amounts of the bets depending on your luck, can change the edge from the house to you. A theorem of probability essentially says you cannot change an unfavorable game into a favorable game by any strategy that just depends on the present and the past history of the game and that does not require the gambler to be infinitely rich.

Let's see that if we *do* allow the gambler to be infinitely rich there is a strategy that allows him to win with probability 1 even if he is playing an unfavorable game. The strategy involves the gambler changing his stakes after each play, that is, the amount of his bets. The method, which seems to have been popular with gamblers over the years, involves doubling the stakes after every loss. Here is how it could work. Suppose the game involves repeated plays where at each play the gambler can bet as much as he wishes, say $\$i$. On that play, he will win or lose $\$i$ with probabilities 1/4 and 3/4, respectively. Assume he starts betting with $i = 1$. The strategy is this: keep playing for \$1 as long as you are winning. If you lose, you bet \$2 on the next play, and keep doubling your bets after each losing play. When you win, you can go back to playing at \$1 again if you choose. If you quit after any play you have won, you will walk away a winner. To see this, we just have to notice that if the gambler starts by losing \$1 and has n consecutive losses, his total losses are

$$1 + 2 + 2^2 + \cdots + 2^n = 2^{n+1} - 1.$$

If he wins on the next play, he gets 2^{n+1}, so he has cancelled out all losses and is ahead by \$1. This game is unfavorable to the gambler in the sense that at each play the expected winnings are negative, yet the gambler has a foolproof method for winning: since with probability 1 the losing streak cannot go on forever, just stop playing after any winning play.

The problem with this strategy is that it requires the gambler (and his adversary) to have infinite capital because it is not known in advance how long his losing streak may last. In practice, the gambler only has finite capital and can go broke; the problem then becomes the classical one of the gambler's ruin (see Chapter 10). In addition, the house has a limit on bets, which means that even if the gambler has the money there will be a certain bound beyond which he will not be allowed to bet.

There have been more sophisticated attempts to turn the tides of fortune. Some more complicated casino games, like Blackjack, require the gambler to make choices. This is in contrast to the simple games we have already described in which only chance operates. When you have several choices, you should, of course, choose the one giving you the greatest probability of winning. Blackjack is played with several decks of cards. The rules require a dealer to deal cards to players who try to reach a total as close to 21 as possible without exceeding it (each card has a number value assigned to it). The dealer, representing the house, is also a participant in the game. A player may have the choice of being dealt another card or cards from the decks during play. Since the probability of winning depends on the distribution of cards remaining in the undealt portion of the decks, so-called "card counters" have tried to remember which cards have appeared in order to have a better idea of the cards remaining; this added knowledge can help them decide between choices and increase their conditional probability of winning. A number of systems have appeared that claim to show how a gambler can get an edge over the house by card counting (for example, see [33]). No one, however, seems to have made a fortune by these schemes. One problem is that you have to be experienced enough to execute a system flawlessly even if there is something to it. The house, moreover, is ready to eject anyone who looks like a counter, and made the job of counting harder by increasing the number of decks used in a game. .

Another assault on the casinos came from a group of physicists, mathematicians, and computer experts described by Bass [1]. They tried to conquer roulette and offered the following argument. Roulette depends on a ball careening over a spinning wheel and is a pure game of chance. On the other hand, according to classical physics, if we know enough about a physical system (large enough, like the roulette system, so that the Heisenberg uncertainty principle need not be taken into account), it should be possible to predict its state in the future. So, in principle, the ball's final resting place on the roulette wheel should be predictable if the basic physical parameters of the ball and roulette wheel and all the various forces at work are known at a certain time at the beginning of the play, and if we have equations at our disposal that can utilize this information. These are big ifs. Moreover, the data would have to be gathered quickly and secretly by watching a particular wheel in action in plays before the bet is made. The group of scientists built computers to fit into a shoe and devised ingenious devices to register the data surreptitiously. They claimed their method worked, but they were plagued by failures of their gadgetry and never did fulfill their dreams. Their few successes might just have been chance fluctuation. A problem with their idea is suggested by the possibility that where the ball lands may be extremely sensitive to initial conditions and round-off error. So although you may have perfect equations of prediction, it may be impossible to supply them with data accurate enough to make solid predictions. The relatively new discipline called the theory of *chaos* may have something to say about this: one of its main ideas is the sensitive dependence of behavior on initial conditions that some physical systems exhibit. If roulette has chaotic behavior, it is inherently unpredictable and it seems we must fall back on the theory of probability.

7.7   Administering a blood test

This problem is given as an exercise in [8]. Suppose a large number N of people have to be given a blood test for a certain disease. The test can be administered in two ways:

a. Each person is given the test separately.

b. The blood samples of i people are pooled, mixed, and then tested. If the test is negative, the single test suffices as a negative test for the i people; if the test is positive, all i people must be tested separately. We do this for groups of i until all N people have been tested.

We assume the test is positive with probability p for each person and that outcomes of the test for individual people are independent (so we have Bernoulli trials with success corresponding to testing positive). Which method of administering the test should we choose, and if we choose (b), how shall we determine the value of i?

The idea behind using the pooled sample test (b) is that testing is expensive, time consuming, and requires equipment that may be scarce, so if (b) results in fewer tests on the average, considerable savings and increased efficiency can be achieved. The Second World War was the stimulus for research into this kind of problem because of the need to perform such tests on large numbers of military personnel. We therefore use the principle that the preferable plan is the one with the smaller expected number of tests.

We will now compute the expected number of tests for each plan. Plan (a) poses no problem; there will be exactly N tests, and probability is not required to figure this out— we have a deterministic situation. Let's look at plan (b), and let X be the total number of blood tests performed for a *single group of i people* in plan (b). Then $X = 1$ when all i people test negative. Each negative test occurs with probability $q = 1 - p$, so by independence all i people test negatively with probability q^i. If the pooled test is positive, then $X = i + 1$ since i further individual tests are now necessary. The probability of a positive pooled test is $1 - q^i$, so

$$EX = 1 \cdot (q^i) + (i + 1) \cdot (1 - q^i) = i - iq^i + 1. \tag{7.2}$$

What we want now is the expected number of total tests for the N people. First, suppose N is an exact multiple of i so that all N people can be broken up into $r = N/i$ groups, where each group has the same expectation as in formula 2. Let T be the total number of tests for the N people using plan (b), and let X_k be the number of tests required for the kth group, for $k = 1, 2, \cdots, r$. Clearly,

$$T = X_1 + X_2 + \cdots + X_r, \tag{7.3}$$

and it seems very reasonable to suppose that from formula 7.3 we might expect

$$ET = EX_1 + EX_2 + \cdots + EX_r. \tag{7.4}$$

Formula 7.4 says that the expectation of the sum of the random variables [given in formula 7.3] is the sum of the expectations of the random variables. This relation turns out to be true, as we will see in the next chapter, for the sum of any random variables. Since each expectation in the sum of formula 7.4 has value equal to the right-hand side of formula 7.2, formula 7.4 yields

$$ET = \frac{N}{i}(i - iq^i + 1) = N(1 - q^i + i^{-1}), \tag{7.5}$$

giving the desired expectation.

Let's concentrate on the term $F = (1 - q^i + i^{-1})$ in formula 7.5. According to our standard of preference, plan (b) is preferable to plan (a) when ET is less than N, and the best version of plan (b) is the one that makes ET as small as possible. ET is proportional to F, so the question becomes: for a given q, find the value of i making F a minimum. In practice q, the probability of a negative test for a person, would be rather large— let's fix ideas by supposing $q = .99$. Now we can see what happens to F as i takes on different values. For $i = 1$ and 2, $F = 1.01$ and $.5199$, respectively. For i extremely large the second and third terms of F are small, close to zero, so F is close to 1. From these observations, it appears that as i increases, F decreases below 1 and then starts to increase to 1. It therefore seems likely that there is a value of i making F as small as possible, and this feeling can be justified by a more rigorous analysis. By playing around with a calculator, it is not very hard to close in on this optimal value of i, which equals 11. This gives an approximate value for ET of $.2N$, that is, with the numbers chosen here, only about 20 percent of the tests given in plan (a) would be performed *on the average*, a savings of 80 percent. Of course, in any particular situation plan (a) might turn out to require fewer tests than plan (b).

There is one remaining matter to dispose of. To make this computation, you recall, we assumed N is a multiple of i so that formula 7.3 could be written down exactly. If N is not a multiple of i, the last group has fewer than i people, and so the sum on the right-hand side of formula 7.3 must be replaced with another sum, where the last term has a different distribution from all the others. But since N is supposed to be large, the right-hand side of formula 7.5 can still be used as a reasonable approximation to the true value of ET even when N is not a multiple of i.

7.8   Exercises for Chapter 7

1. Roll a pair of fair dice repeatedly for 100 independent trials, and let $X=$ the number of times 7 occurs. Express $P(X = 5)$. Express $P(X < 98)$.

2. Let X be your winnings in the game of chuck-a-luck. Find the expectation of X *given that the number you bet on appears at least once*. (Hint: Use the usual formula for the expectation, but use a conditional probability distribution rather than the unconditional distribution.)

3. Here is a finite version of the Petersburg game. As with the classical game, you toss a fair coin until you get a head for the first time, and receive 2^i if this occurs at trial i. In this game, however, you are only allowed a maximum of N trials. If tails occurs at each of the N trials, you win nothing. What entrance fee should you be required to pay in order to make the game fair?

4. Roll a pair of dice until 7 occurs for the first time. What is the expected number of trials until 7 appears? Suppose you play a game by rolling the dice three times. You win if there is at least one 7 and lose otherwise. If you lose, you pay your opponent $3. What should your opponent pay you to ensure that the game is fair?

5. A standard deck of cards contains 4 suits (hearts, diamonds, clubs, and spades) of 13 cards each, numbered cards from 1 to 10, and the jack, queen, and king. The hearts and diamonds are red cards, and the clubs and spades are black. There is thus a total of 52 cards in the deck. You choose two cards at random from the deck. Find the expected number of black cards chosen. Find the expected number of hearts chosen.

6. An urn contains six red balls and four black balls. Three balls are chosen with replacement (each ball chosen is put back into the urn before the next random choice). Find the expected number of red balls selected. Do the same problem, but choose balls without replacement.

8

Baseball Cards, The Law of Large Numbers, and Bad News for Gamblers

> *Bosola.* Didst thou never study the mathematics?
> *Old Lady.* What's that, sir?
> *Bosola.* Why to know the trick how to make a many lines meet
> in one centre.
> John Webster, *The Duchess of Malfi*

8.1 The coupon collector's problem

The Bubbleburst bubble gum company includes a picture card of a famous baseball player in each pack of bubble gum it sells. A complete set of cards consists of ten players. The distribution of the cards is uniform; that is, a pack of gum is just as likely to contain a picture of any one of the ten players. How many packs of bubble gum does someone have to buy, on the average, to get a complete set?

This problem, or a variant, frequently goes under the name of the *coupon collector's problem*. The solution requires learning something about sums, and expectations of sums, of random variables, and about the idea of independence of random variables. Later in this chapter, our discussion of the Law of Large Numbers also needs these ideas.

Very often it is necessary to consider sums of random variables. We have already had to consider such sums, for example, in formula 7.3 of Chapter 7. Here's another, very important, example. Suppose I play the following game with you. A fair coin is tossed repeatedly. If it comes up heads, you give me $1; if tails, I give you $1. Let X_i be my winnings on toss i, so

X_i is 1 or -1 depending on whether a head or tail comes up on toss i. Then $S_n = X_1 + X_2 + \cdots + X_n$ represents my accumulated winnings at the end of n tosses. Is there some neat way to express ES_n in terms of the expectations of the individual variables X_i?

First, we will restrict ourselves to adding up only two discrete random variables, X and Y, which we assume have arbitrary distributions and are defined on the same probability space. Moreover, let us assume each variable has a finite expectation; that is, if an infinite series is needed to define the expectation, the series is summable. Suppose we are interested in $E(X + Y)$. To give such expressions meaning it is not enough to know the probability distributions of X and of Y separately; we must know what is called the *joint* probability distribution of X and Y, the probability weighting of both variables acting together. This joint distribution is the list of the probabilities of the events $\{X = a \text{ and } Y = b\}$ where a and b range over all possible pairings of X and Y. If we have this joint distribution we can define $E(X + Y)$ as

$$E(X + Y) = (a_1 + b_1)P(X = a_1 \text{ and } Y = b_1) \qquad (8.1)$$
$$+ (a_2 + b_2)P(X = a_2 \text{ and } Y = b_2) + \cdots$$

where the addition is over all the possible pairings of X with Y. Now something very nice happens with formula 8.1. No matter what the random variables X and Y are, it can be shown that the result of the addition on the right-hand side always gives $EX + EY$, so that we have the important relation $E(X + Y) = EX + EY$, the expectation of the sum equals the sum of the expectations. Aside from its elegant simplicity and theoretical usefulness, this result is very convenient because EX and EY only depend on the individual distributions of X and Y, not their joint distribution, so the calculation of $E(X + Y)$ becomes easier.

The same ideas hold if we have any finite number, say n, of random variables, each with a finite expectation. The joint distribution and expectation of the sum are defined in a way analogous to the definition for two variables. Again, a computation shows the relation

$$E(X_1 + X_2 + \cdots + X_n) = EX_1 + EX_2 + \cdots + EX_n, \qquad (8.2)$$

which we have already mentioned in section 7.7.

Formulas 8.1 and 8.2 have a more general form. First, notice that for any random variable X we may take any constant c and define a new random variable cX as follows: if X has a value x with probability p, then cX has a value cx with probability p; that is, cX is obtained from X by multiplying all values of X by c and using those with the same probability. It is now very easy to check the rule $EcX = cEX$. So any constant and the E symbol can interchange places. Using formula 8.1 and this rule about constants, it follows that if a and b are any constants, we get

$$E(aX + bY) = aEX + bEY,$$

called the *linearity property* of the expectation. If both a and b are put equal to 1, we get formula 8.1. There is a similar linearity relation generalizing formula 8.2. From the linearity property comes an important observation, that for any random variable X with finite expectation, the random variable $E(X - EX) = 0$. This follows from the linearity property by defining Y equal to the constant EX with probability 1, $a = 1$ and $b = -1$.

As an example of the use of formula 8.2, consider the game described above in which a fair coin is repeatedly tossed and we wish to calculate ES_n, the expectation of the sum of the variables X_i representing the winnings at the ith toss. We have $EX_i = 1 \cdot (1/2) + (-1) \cdot 1/2 = 0$, and the bet on the ith toss is a fair game. S_n, the accumulated winnings after n games, also has expectation 0 by formula 8.2. So the game described by looking at the accumulated winnings after n tosses is also fair for all values of n.

8.2   Indicator variables and the expectation of a binomial variable

An interesting application of formula 8.2 can give us the expectation of a binomial random variable. Suppose there are n Bernoulli trials with success probability p. We now define a sequence of random variables, called *indicator* variables; the idea behind this sequence is very important and indicator variables will be used frequently from now on. For each i between 1 and n, we define the random variable X_i to be 1 if success occurs and 0 if failure occurs on the ith trial. The value of X_i, then, only depends upon what happens on trial i, and it indicates whether or not the trial resulted in success by its value of 1 or 0. Moreover, the sum

$$S_n = X_1 + X_2 + \cdots + X_n$$

of all the indicators is the total number of successes occurring in the n trials, and so S_n has the binomial distribution. Each X_i has the same distribution, and

$$EX_i = 1 \cdot p + 0 \cdot q = p. \tag{8.3}$$

From formulas 8.2 and 8.3 we get

$$ES_n = EX_1 + EX_2 + \cdots + EX_n = p + p + \cdots + p = np, \tag{8.4}$$

and we have obtained the expectation of a binomial random variable with parameters n and p. So if you have a fair coin and toss it 1000 times, the expected number of heads is $(1000)(1/2) = 500$.

The above elegant derivation of the expectation of a binomial random variable depends upon first expressing the total number of successes in n trials as a sum of indicator variables, and then using formula 8.2 to calculate the expectation of this sum. To appreciate the simplicity of this approach, we can try to calculate the expectation of a binomial variable by directly using the definition of expectation: we take each possible value and multiply by the probability of that value, and then add up over all the possibilities. The binomial distribution was derived in Section 7.2; from this we see that we must add up the terms

$$i \cdot \frac{n!}{i!(n-i)!} \cdot p^i q^{(n-i)}$$

from i equal 1 to n. This can be done and, of course, the same answer np falls out, but this method is more computational and less transparent than the slicker indicator approach.

8.3 Independent random variables

We already know what is meant by the independence of events. Roughly speaking, if you have a bunch of events that are independent, any information about what happens with some of the events gives you no information about what can happen with any of the others. The prototypical example to keep in mind illustrating independent events is the case of n Bernoulli trials, where the events {success on trial i} for i equal 1 to n form an independent collection. What, then, are independent *random variables*? Well, it makes intuitive sense to say random variables are independent if they are defined in terms of independent events. For example, in the Bernoulli trial set-up let X_i be the indicators defined for i equal 1 to n as described in the preceding section. These random variables are independent because if we know the values of X_3, X_7, and X_8, say, this is equivalent to knowing whether success or failure occurred at trials 3, 7, and 8. Since the trials are independent, this gives us no information about whether success or failure occurs at any of the other trials, and this is equivalent to saying you have no information about X_i when i is different from 3, 7, or 8. In general, random variables are independent when the events that define them are independent. More precisely, if X and Y are independent, then

$$P(X = a \text{ and } Y = b) = P(X = a) \cdot P(Y = b) \tag{8.5}$$

is true for all possible choices of a and b. This simply says that the events $\{X = a\}$ and $\{Y = b\}$ are independent for all possible choices of a and b. A set of random variables is independent if all finite subsets satisfy a product rule analogous to formula 8.5.

8.4 The coupon collector's problem solved

Finally, we are ready to solve the coupon collector's problem given at the beginning of this chapter. Think of a boy who is collecting the cards and wants a complete set. He buys his first pack of gum and gets his first card. Starting at his second purchase, consider each purchase a Bernoulli trial where there is success if he gets a different picture from the first and failure otherwise. Then p_1, the probability of success, is equal to 9/10. Let X_1 be the waiting time (that is, the number of his purchases) between his first purchase and the purchase that gives him a different picture. After he obtains his second new picture, we can start over and think of each purchase from that point as a Bernoulli trial in which success is defined by getting a picture different from the first two and failure otherwise. Then p_2, the probability of success, is 8/10. Let X_2 be the waiting time between the time (i.e., purchase) he got the second picture and the time he gets a picture different from the first two. We can continue in this way. In general, the purchases between the time the ith distinct picture was obtained and the $(i+1)$st distinct picture is obtained can be considered random variables with success probability $p_i = (10 - i)/10$. We set X_i as the waiting time until he first gets an $(i + 1)$st picture different from the i distinct pictures already obtained. Here i can vary between 1 and 9. The total number T of purchases before he gets the complete set is

$$T = 1 + X_1 + X_2 + \cdots + X_9.$$

The "1" in the formula is due to the first purchase which always yields a picture not already owned. From formula 8.2

$$ET = E1 + EX_1 + EX_2 + \cdots + EX_9.$$

Let's figure out the right-hand side of this formula. First, $E1 = 1$ (a constant can be considered a random variable taking on the constant value with probability 1). Recall Section 7.4 where we discussed the random variable X that measures the time of the first success in a sequence of Bernoulli trials. The expectation of this variable is the reciprocal p^{-1} of the probability p of success at each trial. Each of the variables in the formula above is one of these Bernoulli waiting time variables, so we get $EX_i = 10/(10 - i)$, the reciprocal of the success probability associated with X_i. Therefore,

$$ET = 1 + \frac{10}{9} + \frac{10}{8} + \cdots + 10 = 10\left(1 + \frac{1}{2} + \cdots + \frac{1}{10}\right) \approx 29.25,$$

and the expected number of packages of gum purchased before a complete set is obtained is about 29. It is easy to see that the above argument generalizes to the situation where a complete set consists of n pictures. Therefore, T, the total number of trials until a complete set is obtained, now has the expectation

$$n\left(1 + \frac{1}{2} + \cdots + \frac{1}{n}\right).$$

The coupon collector's problem involves the same interesting idea that we met in discussing the monkey at the typewriter (Section 5.4). This idea concerns doing something until we get a certain result, and then starting afresh. In the present case, we do not know in advance the specific trial when we will get the result (getting a picture distinct from all pictures collected previously)—this trial comes at a *random time*, and then we start afresh unless we are finished. In the case of the monkey, we waited until T trials before we started afresh to try to reproduce Shakespeare's works, unless we were finished. (Instead, we could have waited until the monkey made his first mistake, and then started afresh—this would have given a random time for starting afresh. The mathematics, however, was easier to explain using the fixed time T.) When we start afresh we have an entirely new situation independent of the past. In this way, time is broken up into non-overlapping blocks or segments where events defined in terms of trials on non-overlapping segments are independent. In the present case, this means that after we get a picture different from all predecessors, we can think in terms of a different Bernoulli trial scheme until we first get another picture different from all predecessors, and then we can change again. The random variables X_i are independent: if we know that it took us 12 trials, say, to get our second distinct picture, that is, $X_1 = 12$, it gives no information about the waiting time X_2 for the third distinct picture to appear. As we mentioned in Chapter 5, these intuitively reasonable assertions about independence have rigorous mathematical proofs.

8.5   The Law of Large Numbers

We are finally ready for a discussion of the fundamental theorem called the Law of Large Numbers, mentioned several times in the preceding chapters. There are really a number of such theorems of which the earliest version, a so-called "Weak Law," was limited to sequences of binomial 0-1 variables; this was given by James Bernoulli in the beginning of the eighteenth century. The version we discuss in this section, a "Strong Law," is due to the twentieth-century Russian mathematician A. Kolmogorov.

The general idea behind the Law of Large Numbers can be described as follows: first, think of an infinite sequence $X_1, X_2, \cdots$ of independent random variables taking only non-negative values, all having the same distribution with EX_1 a finite number (the Bernouill trial set-up provides a good example, with X_i the indicator variable taking values 1 and 0 depending on whether the ith trial resulted in success or failure). Since the expectation of a random variable just depends upon its distribution, each X_i has the same expectation, EX_1. Now let $S_n = X_1 + X_2 + \cdots + X_n$ be the sum of the first n X's. S_n is also a random variable, as is the ratio S_n/n, the arithmetic average of the first n X's. We want to concentrate on the averages S_n/n as n gets large. Roughly speaking, the Law of Large

Numbers says that as n gets larger these averages get closer to the constant EX_1 for "almost all" of the sample sequences of $X_1, X_2, \cdots$. That is, "most of the time" the values $x_1, x_2, \cdots$ of the random variables $X_1, X_2, \cdots$ that you get will be such that if you give an arbitrarily small positive number ε, then for all sufficiently large n we have

$$EX_1 - \varepsilon < s_n/n < EX_1 + \varepsilon,$$

where $s_n = x_1 + x_2 + \cdots + x_n$. (Notice that it is important in general to distinguish a random variable X from a value x that it may take on.) Here again, we have that important notion of convergence, with the averages of most sample sequences converging to EX_1; that is, the sequences of these averages get closer and closer to EX_1 as you go farther and farther out in the sequence.

To bring this down to Earth, suppose we have the Bernoulli trial set-up with the indicator functions. Formula 8.3 gives $EX_i = p$, and as we saw in Section 8.2 the sum S_n is the total number of successes in n trials. So in this case the Law of Large Numbers says that "most of the time" for large n the relative frequency of the total number of successes in the n trials will be close to the probability of success, p. In the particular case of applying the model to a fair coin, it means heads should turn up roughly half the time for large n, the approximation to $1/2$ becoming better and better as n increases.

Now what do I mean by "almost all" and "most of the time"? It is impossible to describe this precisely without some advanced mathematics, but the following explanation gives the basic idea (you can skip this without any dire consequences). Recall that we are starting off with an infinite sequence of independent random variables whose outcomes are of interest. The sample space S consists of all possible infinite-tuples $(x_1, x_2, \cdots)$ that can be written down, where x_1 is a value of the random variable X_1, x_2 a value of the random variable X_2, etc. It can be proved that a probability measure P can be defined on this non-discrete sample space using the independence of the random variables and their distribution functions (you might want to glance back at Section 1.4). There are so many sets in S that not all of them can have a probability weight while still maintaining the necessary rules for a useful model. It turns out, however, that if the class of sets is restricted somewhat, P satisfies all the rules on this class and we get a useful model with most of the events of interest to us in this class.

The convergence expressed by the Law of Large Numbers can be written in mathematical notation by

$$P\left(\lim_{n \to \infty} \frac{X_1 + X_2 + \cdots + X_n}{n} = EX_1\right) = 1. \qquad (8.6)$$

The expression "lim" stands for limit and expresses what the sequence S_n/n gets close to. Formula 8.6 can be expressed in words as: the probability that the averages converge to EX_1 is equal to 1. From this it follows that the set of averages not converging to EX_1 is an event of probability 0.

The Law of Large Numbers justifies our intuition that the probability of a repeatable event A can be estimated by the relative frequency of its occurrence for a large number of independent repetitions. The Law therefore allows us to breach the gap between theory and the real world. In the theory, we start off with the probability of an event, p, which may not be explicitly known to us. To get an interpretation in terms of relative frequency, we consider an indicator sequence for the event. To do this, consider independent trials on each of which the event may or may not occur, and let the random variable $X_i = 1$ or 0 depending upon whether the event does or does not occur on the ith trial. So now we have a sequence of Bernoulli trials with probability of success (the event occurs) equal to p. Then the ratio S_n/n is the relative frequency of the occurrence of the event in n trials, and the Law of Large Numbers says that this ratio should be close to $EX_1 = p$ for large n for most sequences of values. So the probability, a theoretical number, can be expected to be close to the relative frequency of the occurrences of the event as long as there are enough trials, or as statisticians like to put it, enough observations. This connection is extremely important for both philosophers and mathematicians. It is a cornerstone of the subject of statistics in which the analysis of data is used to make estimates about unknown parameters. We'll talk about statistics in more detail in Chapter 15.

Here's a concrete example of estimation. Suppose we want to estimate the probability p that a particular component in a certain model of car will fail within one year after purchase. Pick n new cars of this same model *randomly* (we'll get back to this idea—it roughly means any failure in one car should be independent of failures in another) and define the indicator variables $X_i = 1$ or 0 depending upon whether or not the component of car i fails within a year. The discussion above shows that we can get a handle on the unknown probability p by taking data, that is, waiting a year and recording how many cars had components failing within that time. The fraction (i.e., relative frequency) of such failures is a reasonable estimate of the unknown probability p because most such fractions converge to p by the Law of Large Numbers. More has to be said, of course, to tie these general ideas down. For example, it is important to have an idea of how large n must be before we can expect our estimate of p to be within some desired degree of accuracy. There are ways to decide on such n; these are more technical questions we leave unanswered (see Section 15.6, however, for a discussion of a version of this problem with regard to polls and confidence intervals). The main issue right now is to understand what the Law of Large Numbers says, and to get a feeling for its simplicity, beauty, and the ways it can be applied.

In the statement of the Law of Large Numbers, we required that the summand terms X_i be non-negative. This is an unnecessary restriction only imposed to make the initial presentation simple to state. For a more general statement, recall that if x is any number, $|x|$ is known as the *absolute value* of x; it is equal to x if $x \geq 0$ and otherwise it is equal to $-x$. The value x is therefore always the non-negative number obtained by striking out the minus sign that may precede the number, for instance, $|-5|$ is 5. Our more general statement of The Law of Large Numbers allows the summands X_i to have negative values as well as positive ones as long as $E|X_1|$ is a finite number. The restrictions of independence and of the identical distribution of the summands must still hold, and the statement of the theorem is the same. To get an intuitive feel about why $E|X_1|$ should be finite, just recall Section 7.6 where we described a foolproof strategy for the gambler. There were, however, just a few little problems—the gambler had to be infinitely rich and had to be allowed to bet arbitrarily large amounts. In the realistic case, however, of gamblers restricted to finite capital and limited bets, there was no strategy to change an unfavorable game into a favorable one. If we think of $E|X_1|$ as the expected amount won or lost on a bet, then the finiteness of this number can be thought of as a restriction on how much the gambler can bet on each play. From this point of view, the Law of Large Numbers guarantees a nice stable outcome for the average accumulated winnings provided the gambler is not allowed to bet excessive amounts. In the contrary case, the Law of Large Numbers may fail, which should not surprise us too much since, as we saw in Chapter 7, when you don't restrict the amounts of the bets you can get some very strange results.

Later on, we are going to need the fact that the expectation of the averages S_n/n for any n is EX_1. In order to see this, use the linearity property of the expectation (see Section 8.1) to get

$$E\left(\frac{S_n}{n}\right) = E\left(\frac{1}{n}\right) S_n = \frac{1}{n} E(X_1 + X_2 + \cdots + X_n) = \frac{1}{n} \cdot (nEX_1) = EX_1.$$

8.6   The Law of Large Numbers and gambling

Let's see what the Law of Large Numbers says about the game of craps. Suppose X is the gambler's winnings after one game for a $1 bet. In Section 7.4 we saw that $EX = -.014$. Assume the gambler continues to play game after game, and let X_i be the gambler's winnings from game i. Each random variable X_i has the same distribution as X and it is reasonable to assume independence for the X_i. In this case S_n, the sum of the first n X's, are the gambler's accumulated winnings after n games. From formula 8.2 it follows that ES_n, the gambler's expected accumulated winnings after n games, is $\$(-.014)n$, the minus sign indicating an expected *loss*. As the number of games n grows larger, this expectation gets larger without bound. Using the Law of Large Numbers, we can say even more.

Let us notice that in this case

$$E|X_1| = 1 \cdot (.493) + (|-1|) \cdot (.507) = 1,$$

and the conditions are satisfied for the Law of Large Numbers to apply. We get for most sequences of games

$$\frac{S_n}{n} \longrightarrow -.014, \tag{8.7}$$

where the arrow expresses convergence. So the gambler's average accumulated loss after many games is about 1.5 cents, the same as her expected loss after one game. Now, put this way it doesn't sound so bad; 1.5 cents is not such a lot of money. But formula 8.7 is really very depressing news for the gambler. To see this, take any negative number larger than $-.014$, say $-.013$. Consider the event

$$\frac{S_n}{n} < -.013. \tag{8.8}$$

Since the average accumulated winnings are converging to $-.014$, after a sufficiently large number of games these averages are necessarily less than $-.013$ with very large probability, so formula 8.8 is satisfied most of the time. By multiplying both sides of the inequality of formula 8.8 by n, formula 8.8 can be written in the equivalent form

$$S_n < (-.013) \cdot n, \tag{8.9}$$

and this inequality will therefore hold for most sequences of games if the number n of games is large enough. So from formula 8.9 it will follow that if the gambler plays long enough her accumulated winnings will become more and more likely to become more and more negative (since n on the right hand side is becoming larger and larger). That is, her accumulated losses are more and more likely to become large without bound as the number of plays increases. So a small *average* loss translates into a huge *accumulated* loss when the number of games is large, *with probability very close to 1*. This is saying considerably more than just saying the *expected* accumulated loss is huge.

The preceding discussion was only for a $1 bet; if the bet is i, the expectation of loss at each game is $(1.4) \cdot i$ cents, and formulas 8.7–8.9 must be altered by a factor of i. Let's say the gambler bets $10 at each play and plays 100 games. Then formula 8.9 altered by a factor of 10 gives the right-hand side an absolute value of $13; after 1000 games it becomes $130, etc., with the probability that the gambler's accumulated losses exceed these values approaching 1 as the number n of plays increases.

The bad news for the gambler is the good news for the gambler's adversary, who is playing a favorable game at each play. For this adversary, the Law of Large Numbers is a dear friend, assuring the overall taking in of money. A casino is constantly playing, so the number of plays n of any game is indeed going to be large enough so the Law of Large Numbers will be taking effect: with 10,000 plays of craps at $10 a play, the casino will rake in more than $1300 with a probability very close to 1 (since many people can bet at once, the number of plays can get big fast).

8.7   A gambler's fallacy

It is important not to read into the Law of Large Numbers things it does not say. Gamblers, in their desire to win, often misinterpret it. If they have a losing streak at dice, say, they often feel their next throw is more likely to be favorable because, they would say, the "law of averages" guarantees a change of luck eventually. This argument is wrong since each throw of the dice is independent of the previous throws; the dice do not "remember" what happened previously and do not "try" to even the score. At the end of Section 5.1, we briefly mentioned an alternative model of sequences of coin tossing (or dice throwing)—instead of independence the sequences have memory and the probabilities can change depending upon the past history. As we said there, no empirical evidence supports such a view; it is the model of independence that is supported by the data. Of course, there are many other processes where the past history *does* matter and independence is not an appropriate assumption, but the repetitive plays of games of chance are not among these.

What is the basis of the gambler's fallacy? The Law of Large Numbers states that, in the long run, the *averages* in general get close to the expectation. So while you know in dice you cannot keep losing forever (otherwise, with a $1 bet the averages get close to -1 when they should get close to $-.014$), all you know is that there will be times (if you don't go broke first) when you will win if you keep playing. You do not know when these times will be, and you certainly don't know that the next play will be or is likely to be one of these times.

8.8   The variance of a random variable

As we know, the expectation of a random variable is an average of the possible values of the variable. It is a single number giving you an idea of the central tendency of the values. But the expectation gives you no information at all about how the values are *spread out* on the number line around the expectation. For example, if U is defined to be 1 or -1, each with probability $1/2$, U has expectation 0. If V is defined to be 10^6 or -10^6, each with probability $1/2$, V also has expectation 0. But the values of V are much further away from the common expectation 0 than the values of U.

The *variance* $\sigma^2(X)$ of a random variable X is a measurement of how much the values of the variable depart from the expected value. It is defined by the relation

$$\sigma^2(X) = E(X - EX)^2. \qquad (8.10)$$

Notice that the right-hand side of formula 8.10 is just the expected value of the square of the distance of X from its expected value. In other words, for every value of X, we subtract from it EX and then square it; this squaring assures a non-negative value for this term. We then take all of these squares and get the weighted average using the probabilities. Since we are seeking a measure of how the values of the variable deviate from the expectation, you may wonder why we take the *squares* of the distances rather than just the distances themselves in the average. We could, in fact, have defined a measure of dispersion in terms of the distance, but such a quantity would not have mathematical properties as tractable as the variance, and the theory would not be as nice.

For the random variables U and V defined above, we have

$$\sigma^2(U) = (1 - 0)^2 \cdot \frac{1}{2} + (-1 - 0)^2 \cdot \frac{1}{2} = 1,$$

and

$$\sigma^2(V) = (10^6 - 0)^2 \cdot \frac{1}{2} + (-10^6 - 0)^2 \cdot \frac{1}{2} = 10^{12}.$$

The variance of V is much larger than the variance of U, and this reflects the wider dispersion of the values of V from the expectation 0 compared to the values of U from 0.

Formula 8.10 roughly tells us the following: a small variance indicates a relatively high probability of values concentrated close to the expectation, whereas a large variance indicates a relatively high probability of values deviating appreciably from the expectation. For x any value of X, the terms $(x - EX)^2$ are non-negative, so the smallest a variance can be is zero, and this can only happen if the random variable reduces to the constant EX with probability 1. This situation describes the extreme case of a random variable concentrated close to its expectation—all the values coincide with the expectation itself.

Here's an observation that will prove useful to us in a little while. If you take a random variable X with finite expectation, then X and $X - EX$ have the same variance. This follows from the definition of the variance since the expectation of $X - EX$ is 0.

The square root of the variance, called the *standard deviation*, is a quantity much used in statistics. Taking the square root of the variance makes it like a distance function again because the variance is like a squared distance. Statisticians are often interested in the probability of falling within one or two standard deviations of the expectation. We'll come back to this when we discuss the family of normal distributions.

This is the official end of Chapter 8. For those who would like to see a discussion giving some insight into why the Law of Large Numbers is true, I have added an Appendix. Reading the Appendix may be slightly rougher going than what we've been having; the arguments may be a little more complex. You can omit it if you want without any loss of continuity with the rest of the book. You can also forge ahead to try the terrain, and if the going gets too difficult, you can always stop.

8.8.1   Appendix

A careful statement and proof of the Law of Large Numbers requires some advanced mathematics. However, it is possible to give a relatively simple argument showing the *plausibility* of the Law. The basic idea is simple. Let $S_n = X_1 + X_2 + \cdots + X_n$, where the X_i are independent and identically distributed with a finite expectation and variance. We shall calculate the variance of S_n/n and show that as n becomes large this variance converges to 0. The variance being a measure of how far the values of a random variable differ from its expectation, what this indicates is that for large n the probability of the averages S_n/n clustering closer and closer to EX_1 converges to 1. This can be written as: for any fixed positive number ε,

$$\lim_{n\to\infty} P\left(\left|\frac{S_n}{n} - EX_1\right| < \varepsilon\right) = 1. \tag{8.11}$$

This says that the probability of the absolute value of the deviation of the averages from EX_1 becoming less than ε converges to 1 no matter how small ε is. Formula 8.11 expresses what is called *convergence in probability*, which is different from the convergence statement of the Law of Large Numbers as described in Section 8.5 [compare the relation above for convergence in probability to formula 8.6]. Now we are just requiring that the averages are close to EX_1 *with large probability* for any large n. It is perfectly possible in this case for *particular* averages to be close to EX_1 for certain large n and be far for other large n, varying in this way forever, as long as for any large n the probability of the averages that *are* close is very big. In Section 8.5 we were expressing more: that as n gets large most averages (that is, all averages except a set of probability 0) get close to EX_1 and *stay* close to it as n increases. This result about most averages converging, given in formula 5.6 of Chapter 5, is a deeper and more powerful statement than the result embodied in formula 8.11. Formula 8.6 is referred to as the *Strong* Law of Large Numbers, and formula 8.11 as the *Weak* Law. The Strong Law is too hard for us to prove, but the Weak Law is within our grasp, and we are going to indicate the proof of it. This should at least provide us with a little insight as to why the truth of the Strong Law should not be too surprising.

The argument will proceed in several steps. First, we will get a formula for the variance of a sum of independent variables.

8.8.2   The variance of the sum of independent random variables

Can

$$\sigma^2(X_1 + X_2 + \cdots + X_n)$$

be calculated in some easy way in terms of the variances of the individual X terms? We are going to get a nice answer to this question in the case of *independent* random variables. Before we do the computation, though, we need the following observation about independent random variables X and Y, namely, that

$$EXY = EX \cdot EY. \qquad (8.12)$$

In words, the expectation of a product is the product of the expectations. This is analogous to the relation $E(X + Y) = EX + EY$ for the expectation of a sum, but there is a big difference: the relation for the expectation of a sum holds for any two random variables, but formula 8.12 only holds in general for independent ones. The reason this product formula is true can be traced back to the product rule definition for the joint distribution of independent X and Y. Let us convince ourselves of the truth of formula 8.12 in the simple case of two Bernoulli trials, where $X = 1$ or 0 if the first trial results in success or failure and $Y = 1$ or 0 if the second trial results in success or failure. In this case, EX and EY are both $1/2$, and if the terms with product 0 are omitted, we have

$$EXY = 1 \cdot P(X = 1 \text{ and } Y = 1) = P(X = 1)P(Y = 1)$$
$$= \frac{1}{2} \cdot \frac{1}{2} = \frac{1}{4} = EX \cdot EY.$$

So let us accept this product rule and file it away for use in a moment. From the definition of the variance and from formula 8.2 we can write

$$\sigma^2(X_1 + X_2 + \cdots + X_n) = E([X_1 - EX_1] + [X_2 - EX_2] + \cdots + [X_n - EX_n])^2.$$

Now just think about what happens when you square a sum like the one on the right-hand side of this relation. You get the squares of each of the bracketed terms, namely, the terms $(X_i - EX_i)^2$, together with all the cross products, that is, terms like $(X_i - EX_i)(X_j - EX_j)$, for $i \neq j$. After squaring, we must take the expectation of each of the terms in the square and add up. In general, the cross product terms contribute to the sum, but *if the random variables X_i are independent*, something nice happens. Then the expectation of the cross product terms is 0. The reason is: each factor $X_i - EX_i$ has expectation 0, and the random variables $X_i - EX_i$ and $X_j - EX_j$ are independent (subtracting a constant from each of two independent variables again gives two independent variables). So by formula 8.12, we have

$$E(X_i - EX_i)(X_j - EX_j) = E(X_i - EX_i)E(X_j - EX_j) = 0 \cdot 0 = 0.$$

Therefore

$$\sigma^2(X_1+X_2+\cdots+X_n) = E(X_1-EX_1)^2+E(X_2-EX_2)^2+\cdots+E(X_n-EX_n)^2,$$

and since $E(X_i - EX_i)^2 = \sigma^2(X_i)$, this relation may be written

$$\sigma^2(X_1 + X_2 + \cdots + X_n) = \sigma^2(X_1) + \sigma^2(X_2) + \cdots + \sigma^2(X_n)$$

if the X_i are independent. This is the fact we wanted to show: if the random variables are independent, the variance of their sum is the sum of the individual variances. This relation has a similarity to formula 8.2, but formula 8.2 holds for any random variables, whereas the variance formula only holds for independent ones.

8.8.3 The variance of S_n/n

Now let's assume S_n is the sum of the first n terms of a sequence of independent, identically distributed random variables. We will try to compute the variance of the random variable

$$\frac{S_n}{n} - EX_1 = \frac{X_1 + X_2 + \cdots + X_n}{n} - EX_1. \tag{8.13}$$

As we have already seen at the end of Section 8.5, $ES_n/n = EX_1$, so the expectation of the random variable in formula 8.13 is 0 and its variance is the same as the variance of S_n/n. The variable of formula 8.13 can be written

$$\frac{(X_1 - EX_1) + (X_2 - EX_2) + \cdots + (X_n - EX_n)}{n}.$$

This is because each term EX_i is the same as EX_1, and this quantity is subtracted n times and then divided by n. From these observations, it follows that the variance of the variable in formula 8.13 can be written

$$E\left(\frac{(X_1 - EX_1) + (X_2 - EX_2) + \cdots + (X_n - EX_n)}{n}\right)^2$$

$$= E\frac{1}{n^2}((X_1 - EX_1) + (X_2 - EX_2) + \cdots + (X_n - EX_n))^2.$$

From the linearity property of the expectation, the constant $1/n^2$ can go outside the E. Now notice what is left, that is,

$$E((X_1 - EX_1) + (X_2 - EX_2) + \cdots + (X_n - EX_n))^2 \tag{8.14}$$

is the sum of n random variables of the form $X_i - EX_i$. Each of these has expectation 0, and by the definition of variance it follows that the variance of this sum,

$$\sigma^2((X_1 - EX_1) + (X_2 - EX_2) + \cdots + (X_n - EX_n)), \tag{8.15}$$

is exactly what is given in formula 8.14. But since the random variables $X_i - EX_i$ are independent, the preceding section tells us that the variance in formula 8.15 is just the sum of the individual variances. Now the variance, like the expectation, just depends upon the distribution of the random variable, so identically distributed random variables have the same variance. Let $\sigma^2 = \sigma^2(X_1)$ be the common variance of the X's. You recall that we mentioned that subtracting a constant from a random variable produces a random variable with the same variance, so $\sigma^2 = \sigma^2(X_1 - EX_1)$, and formula 8.15 gives $n\sigma^2$ as the expectation of formula 8.14. Remembering the factor $1/n^2$, we get the important result

$$\sigma^2\left(\frac{S_n}{n}\right) = \frac{n\sigma^2}{n^2} = \frac{\sigma^2}{n}. \tag{8.16}$$

Since we are assuming σ^2 is a finite number, the ratio in formula 8.16 converges to 0. Formula 8.16 is the crucial fact leading to formula 8.11. We won't give the mathematical details, which depend upon something called *Chebyshev's inequality*. But intuitively we know that the variance of a random variable measures its spread around its expectation, so that as n increases and the variance of the averages S_n/n converges to 0 [from formula 8.16], the averages cluster more and more about the expectation with probability approaching 1, giving formula 8.11 and convergence in probability, that is, the Weak Law of Large Numbers.

8.9   Exercises for Chapter 8

1. An integer 0 or 1 is picked with probability $1/3$ and $2/3$, respectively, and then a fair die is rolled. Find the expected sum of the first integer chosen and the number rolled on the die.

2. You toss each of three coins. The first is fair, the second has probability $2/3$ of heads, and the third has probability $3/4$ of heads. Find the expected total number of heads for the three tosses.

3. Find the expected number of rolls with a pair of fair dice until 7 comes up ten times.

4. Do exercise 5 of Chapter 7 without using the explicit form of the distributions, but only using symmetry arguments and formulas about expectations. (Hint: If B is the number of black cards chosen in exercise 5, and R is the number of red cards chosen, then $B + R = 2$, and B and R have the same distribution.)

5. Describe how you can use the Law of Large Numbers to estimate (a) the probability of winning by playing black at roulette, (b) the

probability of rolling snake eyes (that is, 2) with a pair of dice, (c) the probabilty of winning at least $1 in chuck-a-luck, (d) the probability of winning if you switch in the car-goat game of Chapter 1.

6. Suppose you have devised a game for which your expected winnings is $0.01. Which, if any, of the following statements are true? (a) If you play a sufficiently large number n of games, you will win $$n$ with probability exceeding .99. (b) If you play a sufficiently large number n of games, you will win $(.009)n$ dollars with probability exceeding .99. Explain your answers.

9

From Traffic to Chocolate Chip Cookies with the Poisson Distribution

"They're falling in a Poisson distribution," says Pointsman in a small voice, as if it was open to challenge.

Thomas Pynchon, *Gravity's Rainbow*

9.1   A traffic problem

Imagine that you are sitting at a pleasant café near a quiet country road, sipping cappuccino and enjoying the scenic countryside. After a while, you notice how little traffic there is on the road. The waiter nods sadly, he says only about five vehicles pass per hour during the afternoon—he wishes the traffic were heavier so business would improve. What is the probability that at least one vehicle will pass in the next 15 minutes?

To try and answer this question, we must get to know an important distribution related to the binomial distribution. It comes about in the following way. Let's say we are counting events happening within a given time, for example, the number of vehicles passing a certain point on a road in 15 minutes, or the number of telephone calls entering a switchboard from nine in the morning until noon. We would like to estimate the probability that exactly k events occur in the given interval of time. It is often helpful in mathematics to draw pictures, so pick up a pen and draw a line segment with left endpoint labelled 0 and right endpoint t (refer to Figure 9.1).

Each point x of the segment represents an instant at which an event did or did not occur (clearly we are idealizing here; an event cannot occur in an infinitesimal instant of time). Now let us draw a little dot at those times

FIGURE 9.1. Dots representing events on a time interval

when an event occurs. We are supposing that the events are discrete in time, so our picture shows an interval with a finite number of the points dotted. Notice that the interval of length t can be subdivided into n subintervals of equal length t/n for any integer n. In order to get a mathematical model that describes quite accurately a variety of physical situations, we are going to make some assumptions:

a. The probability that any very small subinterval of length h contains exactly one of the events is approximately proportional to the length of the subinterval, that is, there is a constant $\lambda > 0$ with this probability $= \lambda h +$ error term, where the error term is extremely small compared to λh for h very small.

b. The probability of two or more events occurring in a subinterval of length h is extremely small compared to λh for h very small.

c. Whatever we know about events happening on any subinterval gives no information about what happens on any subinterval disjoint from the first one.

The assumption (a) is pretty reasonable for small subintervals. As a small subinterval gets a little larger, it becomes more likely that there will be exactly one event in that subinterval. We are assuming the simplest possible form of this increase in probability—that the probability has a linear relationship to the size of the subinterval plus an error term. For small subintervals, moreover, the error term is extremely small compared to the linear relation; that is, the linear relationship becomes more accurate the smaller the subinterval. Assumption (b) is also reasonable given the kind of isolated events under discussion. According to (b), the probability of two or more events happening in a small subinterval is much smaller than the probability of exactly one event occurring. This is what we expect for a finite number of events occurring in a time interval—for tiny intervals it is unlikely to find even a single event, but very much more unlikely to find more than one event. Finally, (c) expresses the independence of events like

{ i_1 events occurred in subinterval 1 }, { i_2 events occurred in subinterval 2 }, $\cdots$, { i_n events occurred in subinterval n }

where the subintervals 1 through n are nonoverlapping, that is, disjoint. The independence assumption is reasonable for many physical processes.

In the traffic case, if some vehicles are observed to pass a point on a road at a certain interval of time, this tells us nothing about what we are likely to observe ten minutes later.

Now take a large value of n and divide up the interval you have into n equal subintervals of length t/n. The three assumptions above will allow us to use the model of Bernoulli trials in order to approximate the probability of k events occurring in the interval. It works like this: each subinterval can be identified with a trial where success means exactly one event has occurred inside it and failure means no event has occurred inside it. But how about a third possibility, that two or more events occur within the subinterval? It turns out that for large n we don't have to worry about this contingency because of assumptions (a) and (b): for small subintervals the probability of two or more events occurring on the subinterval are so small that the event can be safely ignored. The Bernoulli model requires independence of the subintervals, and this follows from assumption (c). So we can think of the k events occurring in the interval of length t as having been produced by n Bernoulli trials in which exactly k subintervals had a single event occur and $n - k$ subintervals had no event occur. The probability of success (a single event occurs in the subinterval) is approximately $\lambda h = \lambda t/n$, because the subinterval has length t/n. From Section 7.2 we can write this as

$$P(\text{exactly } k \text{ events occur in the interval of length } t) \qquad (9.1)$$

$$\approx C_{n,k} \left(\frac{\lambda t}{n}\right)^k \left(1 - \frac{\lambda t}{n}\right)^{n-k}.$$

If you let n increase without bound, it turns out that the right-hand side of formula 9.1 converges to a finite value; that is, it gets closer and closer to some finite number r_k as n increases. Moreover, it follows from our assumptions that as n increases the right-hand side of formula 9.1 should get to be a better and better approximation to the probability on the left-hand side. It therefore makes sense for us to try and *define*

$$P(\text{exactly } k \text{ events occur in the interval of length } t) = r_k.$$

It can be shown that the numbers r_k give a probability distribution, namely, they are non-negative (in fact, positive here) with

$$r_0 + r_1 + r_2 + \cdots = 1,$$

and so we can take this distribution as the exact distribution of the number of events occurring in the interval of length t. This distribution is called a *Poisson* distribution after its discoverer, S.D. Poisson. The quantity λt is called the *parameter* of the distribution.

The expectation of a Poisson distribution (that is, the expectation of a random variable X that has a Poisson distribution) turns out to be its parameter λt. This is not at all surprising. After all, the Poisson distribution was approximated by a sequence of binomial distributions; this is the content of formula 9.1. If there is any justice in the world, the expectations of the binomial distributions on the right-hand side of formula 9.1 should approximate the expectation of the Poisson distribution. In Section 8.2 we saw that the expectation of a binomial variable with parameters n and p is the quantity np. In the present case, the probability of success, p, corresponds to the quantity $\lambda t/n$, so that np corresponds to $n \cdot (\lambda t/n) = \lambda t$. This quantity does not depend upon the particular value of n chosen as n increases, and the expectations of the binomial approximations to the Poisson distribution converge to (actually equal, in this case) λt. Therefore, the expectation of the Poisson distribution should be λt. There is a rigorous proof showing this is indeed the case (so there is some justice in the world). The value λ can be interpreted as the average number of occurrences of the event per unit of time; we call it the *density* of the distribution—to get the average number (or expectation) of events in t units of time, multiply the density λ by t.

It is time now to throw all caution to the wind and show you what the numbers r_k actually look like. To do this, I first have to tell (or remind) you of several facts. One is that if a is any positive real number and b is any real number, then a^b has a well-defined meaning as a real number. What is a real number? We don't have time to get into this in detail; in brief, it consists of the numbers we can identify with all points on a number line. The second thing I have to tell you is that there exists a very famous constant, written e, which, like another famous constant π, turns up all over the place in mathematics. The constant e cannot be expressed as a terminating decimal but is approximately equal to 2.718. With this out of the way, we can write our Poisson distribution as

$$P(\text{exactly } k \text{ events occur in the interval of length } t) \qquad (9.2)$$

$$= r_k = e^{-\lambda t}\frac{(\lambda t)^k}{k!}.$$

To work out the right-hand side we have to know λt and k. According to the remarks above, the powers of e and of λt define real numbers that can be estimated to any desired degree of accuracy with a calculator.

It is apparent from formula 9.2 that there are an infinite number of Poisson distributions. When you specify the parameter λt, you specify a particular one of these distributions. This parameter describes the average number of events occurring in the fixed interval under discussion. The density, λ, is the parameter for the interval of unit length. If we do not know the parameter λt and want to estimate it, we must collect data. In the case of our country scenario, we will assume a Poisson model in which the basic event occurs whenever a vehicle passes the café. To perform the estimate, go back to the café about the same time each day for several months and sit there for about an hour counting the vehicles passing by.

(This doesn't sound like much fun, but we could have a machine do the counting by spreading a cable across the road, where the cable is attached to a counter that records a vehicle every time the cable is compressed.) We count the vehicles about the same time each day because at different times of the day (rush hour versus middle of the night, for instance) a Poisson distribution with a very different density may be in effect. With all the data collected, take the total number of vehicles and divide by the total number of hours and you have an estimate for the average number of vehicles per hour which can be used for the value of λ, the density, since the unit of time is the hour.

We are now ready to attack the problem posed at the beginning of this section. If we accept the waiter's value of an average of five vehicles per hour, then we can take $\lambda = 5$ as the density. Since the unit of time we are adopting is the hour, 15 minutes gives $t = .25$ and there is an average of $\lambda t = (5)(.25) = 1.25$ for 15 minute intervals. Use formula 9.2 now to calculate

P(exactly 0 events occur in the interval of length .25)

$$= e^{-1.25} \frac{(1.25)^0}{0!} = e^{-1.25} \approx .2865.$$

Here we have used the algebraic fact that any number to the power 0 is 1, and that 0! is defined to be 1. The complement of the event "0 events occur" is "at least one event occurs," and so the answer to the problem is $1 - .2865 = .7135$.

9.2   The Poisson as an approximation to the binomial

The Poisson distribution emerged from the assumptions (a), (b), and (c) of Section 9.1. These assumptions, as we said, are often approximately valid in many physical processes. But there is another way to view the Poisson distribution, independently of these assumptions. Mathematically, the Poisson distribution turns out to be a limiting form of binomial distributions; this is what formula 9.1 says—the binomial probabilities on the right-hand side get closer to a Poisson probability as n increases. For large n then, probabilities using the binomial set-up with $p = \lambda t/n$ and number of trials $=n$ can be approximately given by the Poisson distributon. Notice that this binomial set-up has very small p and very large n with $np = \lambda t$ of moderate value. This suggests that whenever we have a binomial situation with p relatively small and n relatively large, the Poisson distribution may be used as an approximation. We don't have to worry about the physical assumptions (a), (b), and (c) in this case; the binomial distribution assumption with large n and small p takes care of everything.

As an example, suppose the probability p that a page of a book has at least one misprint is estimated at about .002, and the binomial model is applied: a trial in this case is a page, with success identified with the appearance of at least one misprint, and the pages are independent. What is the probability that a book of 500 pages has fewer than two pages with misprints? This probability can be calculated exactly using the binomial distribution; it is given by adding up the probabilities of getting exactly zero and exactly one page with misprints (see Section 7.2). We obtain

$$(.002)^0(.998)^{500} + 500(.002)^1(.998)^{499},$$

a rather tedious calculation (even by calculator) which gives us .735959. On the other hand, since p is relatively small and $n = 500$ relatively large, let's use the Poisson distribution to approximate the desired binomial probability. The parameter (which can simply be called λ now) has value $(.002)(500) = 1$. From formula 9.2 we get the sum $e^{-1}(1+1)$, which is .735759. The agreement between the binomial probability and its Poisson approximation is very close.

9.3   Applications of the Poisson distribution

In describing the Poisson distribution, we took an interval of time, but we could have used volume or something else as our physical reference. The important thing is that there be a medium in which the discrete events occur. So, for example, both the distribution of chocolate chips in cookie dough and the distribution of stars in space might be expected to follow an approximate Poisson distribution. Indeed, Poisson random variables turn up rather frequently in life, which is not too surprising because the assumptions (a), (b), and (c) are not very restrictive conditions. Other Poisson or almost Poisson variables include the number of customers arriving at a bank in one hour, the number of α-particles emitted from a radioactive substance in a fixed time, the number of wrong telephone numbers I receive at home in a year, and the number of bottles of mustard sold in a supermarket in a day. In each of these examples, either the assumptions (a), (b), and (c) are physically reasonable or we have a binomial model with p small and n large. Note that if there is a very popular sale item at the supermarket, the number of items of it sold in a day may not be approximately a Poisson variable; in this case, the probability p of a customer purchasing the item may be too large for the approximation to be good.

Some early data showing a close approximation to a Poisson distribution count the number of German soldiers kicked to death by cavalry horses during the years 1875 to 1894. A more recent example is the distribution of V–2 rocket hits in south London during the Second World War. In this latter example, the area under study was divided into 576 areas of equal size, the planar equivalent of the linear subdivision used in Section 9.1. There were a total of 537 hits, giving a value of .9323 for the average number of hits per subdivision, so using the subdivisions as the unit, we get the density $\lambda = .9323$. Define the indicator function X_i to be 0 or 1 depending on whether the ith subdivision is hit or not hit. The expectation of X_i is given by

$$1 \cdot P(\text{subdivision } i \text{ is not hit}) = e^{-.9323} = .3936,$$

using a Poisson distribution with parameter λ for the distribution of hits in a subdivision. The total number of subdivisions not hit, T_0, is the sum of all the X_i, and from formula 8.2 the expectation of T_0 is $(576)(.3936) = 226.71$. If we compare this theoretical number to the actual number of 229 subdivisions, it turns out that they are very close. A similar calculation using a Poisson distribution can be performed to estimate the number of subdivisions hit exactly once, twice, and so on, with results remarkably close to the actual figures observed.

9.4   The Poisson process

In Section 9.1 we considered a line segment of length t, and under the assumptions (a), (b), and (c) we got the result that the number of events occurring in the interval has a Poisson distribution with parameter λt, where λ is some positive constant, the density of the distribution. Since t can vary, we can consider for each t a random variable X_t giving the number of events occurring in an interval of length t. What we have here is a continuum of random variables, one for each specification of t. A collection of random variables like this is called a *random* or *stochastic process*, and this particular one is called a Poisson process. The Poisson process depends upon a single parameter λ which is the parameter of the random variable X_1 of the process.

Stochastic processes are extremely important because very often we are interested in the evolution of some physical system in time. A sequence of random variables X_n is a discrete version of a stochastic process; in a sequence, only integer-valued times are of interest—what happens at roulette, for instance, on a first play, a second play, etc. In the traffic problem, however, we may want to keep a record of the number of vehicles passing the café up to any time t, for all t less than some given time T, for instance. A *sample path* of this Poisson process would be a curve in the $(t, X(t))$ plane over the interval $0 \leq t \leq T$; this would be obtained by plotting how many vehicles $X(t)$ had passed the café up to time t. It seems intuitively reasonable that this curve is non-decreasing and increases only in unit jumps (corresponding to the passing of a vehicle). This turns out to be right, as we will see in Chapter 17, and Fig. 17.2 shows a piece of a typical sample path for a Poisson process. The totality of all possible curves of this type would, in essence, constitute the sample space describing the evolution of the process, which can then be studied using advanced mathematical techniques.

9.5   Exercises for Chapter 9

1. Grandma makes chocolate chip cookies with an average of five chips per cookie. (a) Use the Poisson distribution to estimate the probability of a cookie having six chips. Suppose you ate one of grandma's cookies each week for ten weeks (i.e., the cookies are "independent"). Estimate the following probabilities: (b) at least one cookie had no chips; (c) all cookies had at least one chip; (d) exactly one cookie had three or more chips.

2. A machine stamps out toys in Bernoulli trials where the probability of a defective toy at each trial is .001. Use the Poisson approximation to find the probability of exactly two defective toys in 1000 trials.

3. The number of calls arriving at an office between 9 and 10 AM averages about 30. Use the Poisson distribution to find the probability that on a typical day no calls arrive between 9:45 and 10:00.

4. Customers arrive at a bank in accordance with a Poisson distribution with parameter 40 (based on a unit of one hour). Given that 30 customers arrived in the first hour, find the probability that 60 customers arrive within the first hour and a half.

5. Suppose the probability of an insect's laying r eggs is given by the Poisson distribution with parameter 5, and assume the probability of an egg surviving to develop is p. Also assume the eggs are independent as far as survival is concerned. (a) Write down an expression for the probability of a total of k survivors (your answer should be an infinite series). (b) If we know the insect had at most three eggs, find the probability of exactly one survivor.

10

The Desperate Case of the Gambler's Ruin

Hermann picked a card and placed it on the table, covering
it with a stack of bank notes. It was like a duel. A profound
silence reigned over the gathering.

Chekalinskii started dealing with trembling hands. On his
right showed a queen, on his left an ace.

"The ace has won!" said Hermann and turned his card face
up.

"Your lady has been murdered," said Chekalinskii affably.

Hermann shuddered: indeed, instead of an ace, the queen of
spades lay before him. He could not believe his eyes; he could
not fathom how he could possibly have pulled the wrong card.

Alexander Pushkin, *The Queen of Spades*

10.1 Let's go for a random walk

Suppose you are standing on the number line at 0. With probability p you
go one unit step to the right to 1 and with probability $q = 1 - p$ you go
one unit step to the left to -1. The procedure is then repeated: wherever
you are after the first move, you go one step to the right with probability
p and one step to the left with probability q. If you keep moving according
to this rule, you execute what is called a *random walk* on the integers. Your
position after move n (that is, the integer you are standing on after move
n) is given by the random variable $S_n = X_1 + X_2 + \cdots + X_n$, where the X_i
are independent, taking the values $+1$ and -1 with probabilities p and q,

respectively.

Random walks turn out to be important in a variety of situations. From a gambler's point of view, the variable S_n represents the gambler's accumulated winnings after n plays of a game in which he wins \$1 and loses \$1 with probabilities p and q, respectively. A chemist can use a random walk as a simple model of the random movements of a molecule in a one-dimensional medium (random walks can also be defined in higher dimensions). There are lots of fascinating mathematical questions about the behavior of a random walk. We might, for example, be interested in finding the probability of returning to 0 at some time, or the probability of spending a given fraction of time entirely on one side of 0, or the expected length of time before you arrive at the integer 10, say. We'll discuss some of these questions in Chapter 16, but in this chapter we want to consider a famous old problem concerning our friend the gambler.

10.2   The gambler's ruin problem

We want to investigate the classical problem of the gambler's ruin. This famous question goes like this: a gambler starts off with \$$i$ and plays against an adversary who has \$$(a - i) > 0$. The plays are independent and at each play the gambler wins \$1 with probability p and loses \$1 to the adversary with probability $q = 1 - p$. The play continues until one of the two players goes broke, so that the winning player ends up with \$$a$, the total capital. The question is: what is the probability q_i of the gambler's ultimate ruin, given that the gambler starts with \$$i$?

The gambler's ruin problem involves a random walk, except that initially we must think of the gambler as standing on the integer i instead of at 0. The random walk then starts to evolve until either the gambler arrives at 0 before he reaches a (gambler's ruin) or else arrives at a before he reaches 0 (the adversary's ruin). So this random walk can never leave the interval from 0 to a; the endpoints 0 and a are sometimes referred to as *absorbing barriers* for the walk.

The above statement of the problem seems to indicate that either the gambler or his adversary must win the game. But isn't there another possibility, that the game keeps on going forever without any resolution? Indeed there is, but the solution to the problem will show that the probability of a neverending game is 0, an intuitively reasonable result.

We can solve the gambler's ruin problem completely, but it does require perhaps a slightly longer argument using algebra than we have had up to now. You can always cheat and just look at the answers at the end [formulas 10.6 and 10.8], but I suggest giving it a try to stretch a few algebraic muscles. Here is how we solve the problem of the gambler's ruin.

First, let's see what the sample space looks like. It can be thought of as consisting of (1) the games for which the gambler is ruined: this can be represented as the set of all finite sequences of integers of arbitrary length $(i, y_1, \cdots, y_n, 0)$, where the first entry is i and the last entry is 0, and each entry between the first and last is bigger than 0, less than a, and is one unit apart from its predecessor; (2) the games that the gambler wins (his adversary is ruined): this can be represented by sequences of the type in case (1), except the last entry is a rather than 0; and (3) the neverending games: the set of all infinite sequences with first entry i and all following entries bigger than 0 and less than a.

The probability we seek is the probability of the event in case (1) with last entry 0. The key to finding this probability is the following relation for events:

> {gambler is ruined starting with $\$i$}= {gambler wins $\$1$ and gambler is ruined with $\$(i+1)$} ∪ {gambler loses $\$1$ and gambler is ruined with $\$(i - 1)$}, $1 \leq i \leq a - 1$.

Using this relation and the conditional probability formula, we get

$$q_i = pq_{i+1} + qq_{i-1},$$

and substituting $pq_i + qq_i$ for q_i on the left-hand side, we find the relation

$$pq_i + qq_i = pq_{i+1} + qq_{i-1},$$

which, by rearranging, yields

$$p(q_{i+1} - q_i) = q(q_i - q_{i-1}).$$

Division by p gives

$$q_{i+1} - q_i = \frac{q}{p}(q_i - q_{i-1}), \tag{10.1}$$

true for $1 \leq i \leq a - 1$. Formula 10.1 is an example of a difference equation, a relation between successive differences of the values q_i. Let us define

$q_0 = P$(gambler is ruined with initial capital $\$0$)=1 and
$q_a = P$(gambler is ruined with initial capital a)=0.

These are called the *boundary conditions* for the problem; they specify the situation at the end points 0 and a of the interval, respectively. Here we are saying that if the gambler has $\$0$ he is ruined with certainty, and if he has achieved his goal of reaching a and ruining his adversary, the game is over so his ruin probability is 0.

Using formula 10.1, we can write the following recursion equations which relate each difference to the preceding difference:

$$q_2 - q_1 = \frac{q}{p}(q_1 - q_0) \quad = \quad \frac{q}{p}(q_1 - 1)$$

$$q_3 - q_2 = \frac{q}{p}(q_2 - q_1) \quad = \quad (\frac{q}{p})^2(q_1 - 1)$$

$$\vdots$$

$$q_i - q_{i-1} = (\tfrac{q}{p})^{i-1}(q_1 - 1).$$

At this point perform the following little trick—add up the left-hand sides and the right-hand sides of these recursion equations. On the left, we get what is called a *collapsing sum*: all terms between the first and the ith cancel out. On the right, we have a finite geometric series. Since equalities are added, when we get done both sides are equal. First, let's assume $p \neq 1/2$, so the formula for the sum of a finite geometric series can be used to give

$$q_i - q_1 = \left(\frac{q}{p} + \left(\frac{q}{p}\right)^2 + \cdots + \left(\frac{q}{p}\right)^{i-1} \right)(q_1 - 1) \qquad (10.2)$$

$$= \frac{\frac{q}{p} - (\frac{q}{p})^i}{1 - \frac{q}{p}}(q_1 - 1).$$

[If $p = q = 1/2$, then the right-hand side of formula 10.2 makes no sense because we will be dividing by 0.] Putting $i = a$ in this relation gives us

$$-q_1 = q_a - q_1 = \frac{\frac{q}{p} - (\frac{q}{p})^a}{1 - \frac{q}{p}}(q_1 - 1), \qquad (10.3)$$

which is simply an equation of the form $-q_1 = K(q_1 - 1)$ for K some constant. Solving for q_1 here gives $q_1 = K/(K+1)$; substituting what K is from formula 10.3 leads us to

$$q_1 = \frac{\frac{q}{p} - (\frac{q}{p})^a}{1 - (\frac{q}{p})^a}. \qquad (10.4)$$

Now go back to formula 10.2 and solve for q_i to get

$$q_i = q_1 + \frac{\frac{q}{p} - (\frac{q}{p})^i}{1 - \frac{q}{p}}(q_1 - 1). \qquad (10.5)$$

This gives q_i in terms of q_1, which we've already solved for in formula 10.4. Substituting this into formula 10.5 and using algebra to simplify finally rewards us with the solution in case $p \neq 1/2$:

$$q_i = \frac{(\frac{q}{p})^i - (\frac{q}{p})^a}{1 - (\frac{q}{p})^a}. \qquad (10.6)$$

Turn now to the case $p = q = 1/2$, and observe that the first equation of formula 10.2, which is still valid, reduces to

$$q_i - q_1 = (i - 1)(q_1 - 1). \qquad (10.7)$$

In the case $i = a$, this becomes

$$q_a - q_1 = -q_1 = (a - 1)(q_1 - 1),$$

allowing us to solve for q_1 and get

$$q_1 = \frac{a - 1}{a}.$$

Substitute this relation into formula 10.7 to obtain

$$q_i = 1 - \frac{i}{a} \tag{10.8}$$

when $p = q = 1/2$.

The two formulas 10.6 and 10.8 give us the probability of the gambler's ruin for the cases $p \neq 1/2$ and $p = 1/2$, respectively. The event of the gambler's winning is the same as his adversary's ruin, which can be calculated using the formulas but interchanging p and q and substituting $a - i$ for i. When $p = q = 1/2$, this gives

$$p_i = P(\text{gambler's winning}) = 1 - \frac{a - i}{a},$$

and we immediately check $p_i + q_i = 1$; that is, the game is sure to end after at most a finite number of plays with either the gambler winning or being ruined. If $p \neq 1/2$, we get

$$p_i = \frac{\left(\frac{p}{q}\right)^{a-i} - \left(\frac{p}{q}\right)^a}{1 - \left(\frac{p}{q}\right)^a}.$$

In this case too, a bit of algebra shows $p_i + q_i = 1$.

Here is a scenario where the gambler's ruin model is applicable. Suppose you enter a casino with \$100 and you choose a game such that you will make repeated plays with the same probability of winning \$1 at each play. Your strategy is to keep playing until you go broke or win \$10, whichever happens first. Once you win \$10, you must quit. The gambler's ruin model can be used here as though the casino (the adversary) were to go broke after a loss of \$10. Since the gambler is always playing an unfavorable game at a casino, formula 10.6 with $p < q$ is the appropriate formula in this situation. Put in $i = 100, a = 110$ to get

$$q_i = \frac{\left(\frac{q}{p}\right)^{100} - \left(\frac{q}{p}\right)^{110}}{1 - \left(\frac{q}{p}\right)^{110}}. \tag{10.9}$$

If the gambler's game is craps, in Chapter 6 we found approximate values of .493 and .507 for p and q, respectively, and so q/p is around 1.028. Substitute this into the preceding relation and use a calculator and logarithms to get an approximate value of .253 for q_{100}. The gambler starting with $100 will therefore be ruined about one time in four before he wins $10 at craps.

Since the probabilities p and q in craps are both very close to .5, it is interesting to compare the result we just found with what we would get from formula 10.8 when the game is fair. From that formula, we find a probability of about .091 of ruin for a gambler starting with $100 and playing until ruin or until he realizes a profit of $10. This is less than one time in ten. This example shows that even a small shift in probabilities converting a fair game into an unfavorable one for the gambler can increase the chances of ruin significantly.

10.3   Bold play or timid play?

Now let us consider the interesting question: can the gambler improve his chances of winning by changing the stakes, that is, changing the amount he bets. Assume he keeps the same stakes throughout the game but he has a choice initially. In the case of our gambler who has $100 and wants to win $10, suppose he can bet at $1, $2, $5, or $10 stakes. Which should he choose? Changing the stakes to $2 means the unit of the mathematical model has been changed, so the formulas we have must be changed to reflect that fact. If the unit is now $2 = 1$ chip, in terms of that unit the gambler's initial $100 must be halved to 50 chips, and the amount he has when he quits must be halved to 55 chips. The values p and q remain the same, and formula 10.9 becomes

$$q_i = \frac{\left(\frac{q}{p}\right)^{50} - \left(\frac{q}{p}\right)^{55}}{1 - \left(\frac{q}{p}\right)^{55}}.$$

If 1.028 is again put in for the value of q/p in craps, the answer now is about .165. So the gambler was able to reduce the probability of ruin from .253 by doubling the stakes from $1 to $2. At $5 stakes, the unit is $5 and this corresponds to an initial capital of 20 chips and a goal of quitting at 22 chips. Using a version of formula 10.9 altered appropriately, the probability of ruin is now seen to be about .118. Finally, at $10 stakes the ruin probability calculated in the same way turns out to be about .104. So if the gambler plays at $10 stakes rather than $1 stakes, he is able to reduce his chances of ruin from about one in four to about one in ten.

The above calculations show that in the present example the gambler does best to play at the highest stakes available to him to achieve his limited goal. An algebraic analysis of formulas 10.6 and 10.8 (which we omit) proves that this is a general phenomenon, namely, *in the gambler's ruin model, bold play (that is, betting at the highest possible stakes) is always the best strategy for a gambler playing an unfavorable game ($p < q$).* Conversely, if the game is favorable for the gambler, he should play at the lowest possible stakes.

One way to try to get insight into this fascinating result is by observing that low stakes means the game drags out longer on average and the Law of Large Numbers can take its toll on the player bucking the odds. With larger stakes, on the other hand, the games are shorter so there is less time for damage to be done. This argument may or may not seem reasonable to you, but if it doesn't, don't despair—just be happy we have mathematics available to give us rigorous proofs.

As with all of mathematics, it is important not to misread or misinterpret the results. If the gambler plays an unfavorable game, we already know from Chapter 7 that changing the stakes cannot convert this into a favorable game. What the gambler can do, however, is institute a kind of damage control—although he can't get the odds in his favor, he can minimize his probability of ruin if he has limited aspirations, that is, by quitting when he wins a certain amount if he isn't ruined first. His strategy for damage control is to play boldly, at the highest stakes possible for his goal.

10.4   Exercises for Chapter 10

1. You are playing craps at even money, that is, if you win a game you receive $1 or else you lose $1. You start with $3 and decide to play crap games until you either go broke or accumulate $6. Find the probability of your ruin. If you can change the stakes, what is your best strategy for play?

2. Do exercise 1 for the game of roulette, where you are betting on red at each play.

3. Suppose you are playing the classical game (of Section 10.2) with $p = q = .5$, and you start with $\$i$ and will quit when you attain $\$2i$. Assuming you can choose the stakes (in dollar units), show that bold play is no more advantageous to you than timid play. How should you gamble if p is changed to (a) .499, (b) .501.

4. Ginger and Fred are playing the following game: Fred gives Ginger $5 with probability 1/5 and Ginger gives Fred $5 with probability 4/5. Ginger enters the game with $5, and the game will continue until Ginger goes broke or her fortune reaches $15. Ginger knows the game is unfavorable to her, but Fred assures her that if she is ruined he will dance with her all night. (a) Without using the formulas, calculate directly Ginger's probability of ruin. (Hint: Ginger can only be ruined at the first play, the third play, the fifth play, etc. Calculate each of these ruin probabilities and get an infinite series, which you should be able to add up.) (b) Now use a formula from Section 10.2 to get the probability of Ginger's ruin. Your answer should, of course, agree with the answer you got in (a).

5. Suppose a game is played similar to the classical game of Section 10.2 except that the gambler, starting with $\$i$, can either win a fixed $\$s \geq 1$ or lose \$1 at each play. Assume the gambler quits either upon ruin or the first time a fortune of $\$a$ or greater is attained, $a > i$. Let v_i be the probability of the gambler's ruin in this game and q_i the probability of the gambler's ruin in the classical game. Give a heuristic (intuitive) argument without any calculation showing why the inequality $v_i \leq q_i$ is plausible.

11

Breaking Sticks, Tossing Needles, and More: Probability on Continuous Sample Spaces

All things flow.
Heraclitus

11.1 Choosing a number at random from an interval

Up to now, we have been working with discrete sample spaces and discrete random variables (see Chapters 1 and 7). There are problems, however, for which a discrete sample space is not appropriate because there are just too many possible outcomes. Suppose, for instance, that I want to choose a number on the interval between 0 and 1 "at random." Ignoring for a moment exactly what I mean by the term "at random" in this context, we note that the number chosen can be any value on the interval, so there is a continuum of possible values. This continuum has so many numbers in it that they cannot all be counted off using the positive integers; for this reason the sample space of this interval is not discrete but is what is called a *continuous* sample space.

The first important problem to deal with in the case of continuous sample spaces is to determine what the basic events of interest should be. In the discrete case, each outcome was usually assigned a positive probability. To determine the probability of a more complicated event, we just added up the probabilities of all the outcomes contained in the event. Now matters are more complicated. In a continuous sample space like an interval of numbers, it is impossible for all outcomes, that is, all numbers in the interval, to be given a positive probability. If you try to do this, you cannot preserve the rule that requires the sum of all the probabilities to be equal to 1 (the sum won't even be finite). This suggests that the individual outcomes can no longer be the basic building blocks used in calculating the probabilities of events, as they were in the discrete case. We need a new idea.

The key concept to get us going in the continuous case is to consider *subintervals* of the interval sample space to be the new building blocks for calculating the probabilities of events. If, for example, the sample space is what we call the unit interval, the interval 0 to 1, and we set X = a value selected from this interval, then we are no longer interested in outcomes of the kind $\{X = a\}$ but rather in events of the kind $\{a < X < b\}$, where $0 \leq a < b \leq 1$. It is these latter interval events that will in general have positive probability and from which we will be able to calculate the probabilities of more complicated events. The probability rules we learned in previous chapters still hold, so, for instance, if $P(0 < X < 1/4) = p_1$ and $P(3/4 < X < 1) = p_2$, then the probability that X is in either the interval 0 to 1/4 or 3/4 to 1 is $p_1 + p_2$.

Now that we know our basic events are intervals, we can ask what it means to choose a value at random from the unit interval. If we are choosing a number at random in the interval, then what seems to be called for is a notion whereby each point has the same probability of being selected as any other point. As we saw above, it would be impossible to give such a probability distribution. Since the basic events of interest are no longer individual points, it makes sense to search for a continuous analog of the discrete uniform distribution. Instead of having each outcome with the same probability as any other outcome, we will require of a continuous uniform distribution that any subinterval will have the same probability as any other subinterval *of the same length*. This means that under the assumption of equal likelihood the probability of a subinterval is proportional to its length. For the unit interval, the probability of a subinterval *is* its length; for the interval 0 to 20, the probability of an interval is its length divided by 20. In general, we say that a random variable X defined on an interval I has the *uniform* distribution if the probability that X lies in any subinterval U is equal to the length of U divided by the length of I. We can show the probability of any outcome $\{X = a\}$ must be 0. The event $\{X = a\}$ is included in an interval I_m of arbitrarily small length m, and by the laws of probability, $P(X = a) \leq P(I_m) = m$. Since m is a positive number as small as you like, it follows that $P(X = a) = 0$.

There is a nice pictorial way to represent the probabilities of intervals for continuous sample spaces. We are going to present this now for the very simple case of a uniform distribution. Suppose we take the interval I from 0 to 20 lying on the x axis in the x-y plane (see Fig. 11.1). Let us draw the line L joining the points $(0,1/20)$ and $(20,1/20)$; L lies parallel to I a distance 1/20 units above I. Now if we are given any subinterval J in I, the probability of J may be obtained by getting the area of the region above J and bounded by L; indeed, this area is just a rectangle with length the length of the subinterval J and height = 1/20, and the area, namely, the product of these quantities, gives the uniform probability. Let us now

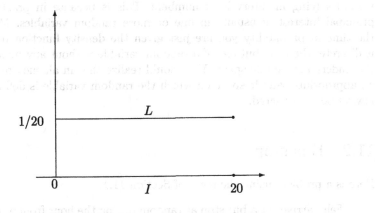

FIGURE 11.1. Density for a uniform distribution (vertical units enlarged)

define the following function

$$f(x) = 1/20, \ 0 < x < 20; f(x) = 0 \text{ elsewhere.}$$

The function $f(x)$ is called the *density function* of this uniform distribution. The density function assigns weight to the individual points of the number line. These weights are not themselves probabilities, but by determining the area above an interval they determine the probability of that interval. In the present case of the uniform distribution, the density $f(x)$ restricted to I is the constant $1/20$ over I, indicating that each point of I is given the same positive weight. The graph of $f(x)$ restricted to I is the line L. But once we leave I the density is zero—the graph of $f(x)$ jumps down to the x-axis outside I. So if we take the interval K from 30 to 40, say, $f(x)$ assigns K probability 0 since the area under $f(x) = 0$ and above K is 0.

In general, a distribution for a continuous random variable X is a probability assignment $P(J)$ to intervals J of the real number line giving the probability that X lies in the interval J. Usually, the way to specify this distribution is to give a density function for the distribution, that is, a non-negative function or curve $f(x)$ defined in the x-y plane such that the probability of any interval J is obtained by finding the area of the region of the plane under $f(x)$ and above J. Moreover, the total area under the density above the entire real line is always 1. An argument similar to the one above for the uniform density shows $P(X = a) = 0$ for any point a when X has a density function. The cumulative distribution function (or just *distribution function*) of a random variable X is the function $F(t) = P(X < t)$. If X has a density function $f(x)$, then $F(t)$ is the area under $f(x)$ over the interval $x < t$.

As you can see from this description, we are shifting the emphasis away from talking about sample spaces and more towards talking about random variables lying in intervals of numbers. This is because in practice the principal interest is usually in one or more random variables. Much of the time in probability you are just given the density function (or, if it is discrete, the distribution) of a random variable without any mention of the underlying sample space. You should realize that in all such problems an appropriate sample space on which the random variable is defined can always be constructed.

11.2   Bus stop

Here is a problem using the ideas of Section 11.1.

> Felix arrives at a bus stop at random during the hour from 6 to 7 pm. If buses leave the stop for Felix's destination at 6, 6:30, and 7, what is the probability that Felix will have to wait more than ten minutes for a bus?

As sample space let's take the time interval S from 0 to 60 minutes and define $X = $ the instant Felix arrives at the bus stop measured in minutes after 6 o'clock. Moreover, the "at random" phrase in the wording tells us that we should use the uniform distribution for X on the interval: Felix is just as likely to arrive in any time intervals of the same length. Now let's see how to solve the problem. Felix has to wait more than ten minutes for a bus if he arrives between 6 and 6:20 or between 6:30 and 6:50 (note that Felix misses the 6 o'clock bus with probability 1 since the probability that he arrives at exactly 6 o'clock has probability 0). So the event "Felix waits more than ten minutes" can be expressed as the union of the events $I = I_1 \cup I_2$, where I_1 and I_2 denote the time subintervals 6 to 6:20 and 6:30 to 6:50, respectively. From the definition of the uniform distribution, both I_1 and I_2 have probability 20/60=1/3, so the union I of these non-overlapping events has probability 2/3. Therefore, Felix will have to wait more than ten minutes for a bus every two out of three times on the average.

11.3   The expectation of a continuous random variable

In previous chapters, we restricted ourselves to discrete sample spaces and random variables. For a discrete variable X, we defined (in Chapter 7) EX, the expectation of X. The value of EX is calculated by taking each of the discrete values of X, multiplying by the probability that X assumes this value, and then adding up over all the possibilities. For continuous random variables (that is, random variables taking on a continuum of values) this definition of EX is no longer possible in general because the individual

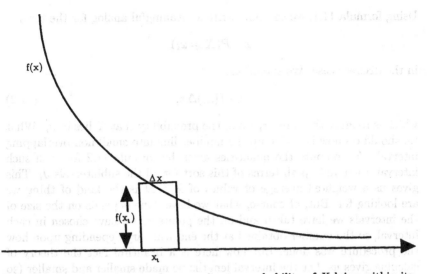

FIGURE 11.2. Area of rectangle approximates probability of X lying within its base

outcomes usually have probability zero, and our basic building blocks are now intervals rather than individual outcomes. Is there a way to extend the notion of expectation to continuous random variables?

The answer to this question is a resounding "yes," and this is just the type of problem for which calculus was invented pretty much simultaneously by Isaac Newton in England and Gottfried Wilhelm Leibniz in Germany in the seventeenth century. What I can do here is try to give you a rough feeling for the ideas involved; they are very beautiful and intuitively appealing. The expectation of X in the discrete case is a weighted average of the discrete values of X. To get the expectation, we added up terms of the form $x_i \cdot P(X = x_i)$. In the continuous case, however, we know this product will be 0, so what we do is quite reasonable—we try to get intervals that have positive probability into the act. Remember that now there is a density function $f(x)$ defined on the number line. Think of this number line divided into intervals that all have the same small length, which we can write Δx. Now if $f(x)$ is a function that isn't too weird, its variation over each tiny interval can't be too great, so given such an interval, say J_i, we can choose any point x_i inside it and consider the term $f(x_i)\,\Delta x$. This term is the area of a little rectangle with height $f(x_i)$ and width Δx and is an approximation to the area bounded by the curve $f(x)$ and above J_i (see Fig. 11.2). Now we recall that the area under a density and above an interval is the probability that the random variable X belonging to that density lies in that interval. This means that

$$f(x_i)\,\Delta x \approx P(X \text{ lies in } J_i). \tag{11.1}$$

Using formula 11.1, we can now write a meaningful analog for the terms

$$x_i \cdot P(X = x_i)$$

in the discrete case. We should use

$$x_i \cdot f(x_i)\Delta x, \tag{11.2}$$

which is roughly the value x_i times the probability that X lies in J_i. What we should do now is divide up the number line into small non-overlapping intervals J_i, compute the quantities given by formula 11.2 for each such interval, then add up all terms of this sort for all the subintervals J_i. This gives us a weighted average of values of X that is the kind of thing we are looking for. But, of course, what we have done depends on the size of the intervals we have taken and on the points x_i we have chosen in each interval, so the number obtained at the end will vary depending upon how the procedure was done. But now here is a beautiful fact the theory of calculus gives us. Let the interval lengths be made smaller and smaller (so the number of intervals is increasing without bound) and consider the sum of the terms in formula 11.2. Each such term is getting close to zero since Δx is getting close to zero, but the number of terms in the sum is growing without bound. If $f(x)$ is a reasonably "well-behaved" density, then the sums of the terms in formula 11.2 approach a unique limiting value (i.e., they get closer and closer to something). This limiting value we define to be the expectation of X. The methods of calculus also give us nice methods to calculate these expectations. Of course, just as in the discrete case, EX may fail to exist. This happens if the density is not sufficiently restricted. If X is uniformly distributed on the interval from a to b, where $a < b$, then EX turns out to have the value $(a + b)/2$, a very reasonable result. So in the bus problem above, we can say that Felix's expected time of arrival at the bus stop is at 6:30.

11.4   Normal numbers

An interesting application of the uniform distribution on the unit interval concerns the so-called "normal" numbers. (Mathematicians have a habit of naming objects normal whenever they display properties that seem to be typical, desirable, or just pleasant.) To define what a normal number is, think of expanding each real number between 0 and 1 in the standard decimal system. Each finite decimal (like, e.g., .135) has two decimal expansions, one of which is non-terminating with a string of 9's (.135 = .134999···). Except for these, each real number in the interval has a unique decimal expansion. If we agree always to pick the non-terminating expansion in the case of the finite decimals, then each real number in the interval corresponds uniquely to an infinite decimal $.x_1x_2x_3\cdots$. Let k be one of the digits 0 through 9. For any positive integer n, we can define the random variables $Z_n^k(x)$ for $0 < x < 1$ by

$Z_n^k = $ (number of the first n digits in the decimal expansion
    of x which are equal to k)/n

The variables Z_n^k simply give the relative frequency of the appearance of the digit k in the first n digits of the decimal expansion of the number x. Then x is called *simply normal to the base 10*, if, as n gets large without bound, Z_n^k converges to $1/10$, that is, it gets closer and closer to $1/10$ as n increases, for each of the ten values of $k = 0, 1, \cdots, 9$. We are going to call "simply normal to the base 10" just "normal" to simplify the discussion, but you should be aware that in the mathematical literature the nomenclature "normal" without qualifiers is often used to mean that x is not only simply normal to the base 10, but has other related properties that are admirable but which we will not go into here.

We can express normality of x in mathematical notation as

$$\lim_{n \to \infty} Z_n^k = \frac{1}{10}, \qquad k = 0, 1, \cdots, 9.$$

In plain words, a normal number x has the nice intuitively appealing property that, on the average, each one of the ten digits 0 through 9 appears $1/10$ of the time in the decimal expansion of x.

By considering finite or repeating decimals, we can see immediately lots of non-normal numbers (e.g., $.1212\cdots$). The more interesting question is how we can find normal numbers or, more fundamentally, whether normal numbers even exist. I am now going to outline an argument using the Law of Large Numbers showing that there are loads of normal numbers in the interval 0 to 1. In fact, "most" of the numbers in the interval are normal. To do this, we consider the uniform distribution on this interval. For any x in the interval, let $.x_1 x_2 \cdots$ be its decimal expansion. First, let's concentrate on the digit 0. Define the random variable U_i to be the indicator variable for the digit 0 at the ith place in the decimal expansion of x; that is, $U_i = 1$ if the ith digit in the expansion of x is 0 and $U_i = 0$ otherwise. Then we have

$$\frac{U_1 + U_2 + \cdots + U_n}{n} = Z_n^0.$$

Now, it is a fascinating fact that because we are using the uniform distribution it is possible to prove that the variables U_i are independent. To get a feeling about why this should be so, let's think for a moment about how the decimal expansion of a number in the unit interval determines its location as a geometric point of that interval. Divide the unit interval up into ten equal subintervals, what we will call the subdivision of first order. The first digit x_1 of the decimal expansion of x locates x in subinterval $(x_1 + 1)$. So if $x_1 = 0$, x lies in the first subinterval from 0 to .1, if $x_1 = 1$, x lies in the second subinterval from .1 to .2, etc. Now divide each of the ten subintervals themselves into ten equal subintervals. We can call this second subdivision the subdivision of second order. Suppose we are given the first two digits x_1 and x_2 of the decimal expansion of x. These two digits locate x in the subinterval $(x_2 + 1)$ of second order within the subinterval $(x_1 + 1)$ of first order. Proceeding in this way, successively divide each subinterval of order i into ten equal parts to get subintervals of order $i + 1$. Then successive digits of x locate x on smaller and smaller subintervals so that when *all* digits of x are given (theoretically), x is uniquely determined as a point on the interval.

Return now to the variables U_i. Let's try to calculate $P(U_1 = 1)$, where P represents probability using the uniform distribution on the unit interval. This is just the probability that the first digit in the decimal expansion of a number is 0; that is to say, the number lies in the subinterval from 0 to .1. The uniform distribution assigns the probability we seek to be the length of this subinterval, .1. How about $P(U_2 = 1)$? For the second digit to be 0, the number must lie in the first subinterval of second order. Each such subdivision has length .01, but since the event $\{U_2 = 1\}$ imposes no restriction on which of the subintervals of first order the number lies in, it could be in any of the ten disjoint possibilities, so $P(U_2 = 1) = (10)(.01) = .1$. Now calculate

$$P(U_1 = 1 \text{ and } U_2 = 1),$$

the probability that the first two digits are both 0. This event requires that x lies in the first subinterval of second order within the first subinterval of first order, so the probability equals .01. From this we see

$$P(U_1 = 1 \text{ and } U_2 = 1) = .01 = (.1)(.1) = P(U_1 = 1) \cdot P(U_2 = 1),$$

which shows that the events $\{U_1 = 1\}$ and $\{U_2 = 1\}$ are independent. By going forward with this argument, we can show that the sequence of the U_i variables is independent as claimed.

Not only are the U_i independent, but they have the same distribution: $P(U_i = 1) = .1, P(U_i = 0) = .9$. But now the stage is set for the Law of Large Numbers. According to that theorem, we have

$$P\left(\lim_{n \to \infty} \frac{U_1 + U_2 + \cdots U_n}{n} = EU_1\right) = 1;$$

that is, the set of all points in the unit interval whose decimal expansions average out (in the limit) to EU_1 has probability 1, and since $EU_1 = (1)(.1) + (0)(.9) = .1$, this formula just says that most numbers have decimal expansions in which the digit 0 occurs, on the average, .1 of the time. Exactly the same argument can be given for each of the ten digits, and it follows that most numbers are normal in the sense that the normal numbers have probability 1 when probabilities are calculated using the uniform distribution.

What does this set of normal numbers look like? Well, it cannot be a subinterval because every subinterval contains lots of finite decimals that are not normal. Unfortunately, when we deal with continuous sample spaces, there are so many points and so many sets that can be formed from them that the mathematics of the situation can get quite delicate and the sets can get quite complicated. The basic idea is that starting with unions and intersections of intervals we can get a whole bunch of sets for which probability statements can be defined in a meaningful way—these sets are called *measurable*. The probability distribution given for intervals, for instance, the uniform distribution, can be extended to a probability function on all the measurable sets such that the usual rules for probabilities that we had in the discrete case still hold or have clear analogs. All this takes a bit of serious mathematics, which we don't have to worry about here. The important thing for us is that the set of normal numbers, while not as simple as an interval or the union of intervals, is a measurable set with probability 1. If you are trying to visualize it geometrically, you can think of the unit interval with a lot of holes in it where numbers are missing. Every subinterval of the unit interval has lots of these holes—we say the holes are *everywhere dense*. Yet the missing numbers don't add up to very much in the sense that the totality of them only has probability zero. So the normal numbers constitute most of the numbers in the unit interval, in fact, almost all of them in a technical mathematical sense.

We have just seen that there are loads of normal numbers in the unit interval. In fact, they are so numerous that the set of them has probability 1. Yet there is something apparently strange here when you think a little further. In spite of the multitude of normal numbers, it is not particularly easy to produce examples of them. Here is one:

$$.01234567890123456789 \cdots$$

where the pattern of the ten digits is repeated infinitely. But presented with a number x in the unit interval, unless it is a very special kind like a finite or repeating decimal, nobody knows how to determine whether x is normal. Nobody knows, for example, whether the decimal part of the number π is normal even though π has been calculated to perhaps as many as a billion decimal places. The reason why identification of normal numbers is so difficult is pretty clear: when you say a number is normal, you are saying something about the *entire* decimal expansion of the number. In general, it seems difficult to understand the patterns in the complete decimal expansions of numbers. No *partial* decimal expansion, no matter how long, can ever tell us by itself whether a number is normal since the limit statement in the definition of normal numbers does not depend on any finite number of decimal places. So we have a kind of paradoxical situation that arises from time to time in mathematics—it is relatively easy to prove certain objects exist but hard to exhibit examples of the objects. We have proved that normal numbers are all over the place, and that if we choose a number x at random from the unit interval it will almost certainly be normal, but unless we are given some special relations among the digits, there is no general way known at present to verify that any particular x is, in fact, normal.

11.5 Bertrand's paradox

In 1889, L.F. Bertrand presented the following problem:

> Suppose you have an equilateral triangle inscribed in a circle. Choose a chord of the circle at random. What is the probability that the length of the chord is greater than the side of the inscribed triangle?

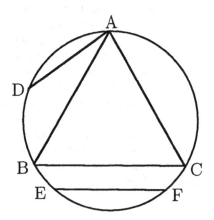

FIGURE 11.3. Equilateral triangle ABC inscribed in a circle with chords AD and EF

To understand this problem, you will have to recall a few facts and a little terminology from plane geometry (see Fig. 11.3). A *chord* of a circle is a straight line segment whose endpoints lie on the circumference of the circle. An equilateral triangle has all three sides of the same length and all three angles equal to 60 degrees. A triangle is inscribed in the circle if its three vertices lie on the circumference of the circle. The problem is sometimes called *Bertrand's paradox* because at least three different answers seem possible by apparently impeccable arguments. The key to understanding the paradox is to realize that choosing a chord at random is not precisely defined. There are several ways to interpret this random choice, and each of these yields a different answer to the problem.

First solution: Draw a radius of the circle perpendicular to a side of the triangle. Choosing a chord at random can be interpreted as choosing a point Q on this radius according to the uniform distribution and then drawing the chord through Q perpendicular to the radius. It is clear that the chord thus drawn is larger than a side of the triangle if Q lies within the triangle. To calculate the probability that Q lies inside the triangle, we only need to find the distance from the center of the circle to the side of the triangle. Elementary geometry shows this is $r/2$, where r is the radius of the circle. It follows that the probability we seek is $(r/2)/r = 1/2$.

Second solution: Draw a tangent T to the circle at a vertex V of the triangle. Consider all chords of the circle that have V as an endpoint. Any such chord makes an angle with T between 0 and 180 degrees. Conversely, given such an angle a unique chord is determined. Choosing a chord at random can be interpreted as choosing an angle between 0 and 180 degrees according to the uniform distribution. The chord is greater than a side of the triangle whenever the chord lies partially within the triangle. Elementary geometry shows this happens when the angle between T and the chord lies between 60 and 120 degrees. The probability is therefore $(120 - 60)/180 = 1/3$.

The reader is encouraged to find another interpretation of choosing a chord at random giving yet a third answer.

11.6   When do we have a triangle?

Here is a cute little problem dealing with two independent uniform distributions (see Fig. 11.4).

> Suppose you are given the unit interval. You choose a first point X at random and then, independently, you choose a second point Y at random. By "breaking" the interval at the points X and Y, three line segments are formed. What is the probability that a triangle can be formed from these three segments?

This problem is not so easy until you look at it the right way. Notice that choosing the two points X and Y can be considered as choosing a single point (X, Y) in the unit square of the plane (the unit square has vertices $(0,0)$, $(1,0)$, $(0,1)$ and $(1,1)$). Since the points are chosen independently and uniformly, if $0 \leq a < b \leq 1$ and $0 \leq c < d \leq 1$

$$P(a < X < b \text{ and } c < Y < d) =$$
$$P(a < X < b) \cdot P(c < Y < d) = (b - a)(d - c),$$

which simply says that the probability of (X, Y) lying in a rectangle within the unit square is the area of the rectangle. In general, the probability that (X, Y) lies in any region of the unit square is just the area of this region. The next thing to do is to try and find conditions on X and Y

FIGURE 11.4. Unit interval divided into three pieces by points X and Y

implying that the three segments formed determine a triangle. First, let us assume case 1: $0 < X < Y < 1$. Then the segments formed have lengths $X, Y - X$, and $1 - Y$. The criterion for the existence of a triangle to be constructible from three segments is very simple: the sum of the lengths of any two segments must be larger than the third (this is a reflection of the geometric fact that the shortest distance between two points is a straight line). Taking the three lengths and imposing these conditions, we get

$$
\begin{aligned}
X + (Y - X) &> 1 - Y, \\
(Y - X) + (1 - Y) &> X, \\
(1 - Y) + X &> Y - X,
\end{aligned}
$$

which simplifies to the three pleasant inequalities $X < .5, Y > .5, Y - X < .5$. It is not hard to see (if you know just a little bit about plotting points in a plane) that the set of (X, Y) satisfying all three inequalities forms the inside of the triangle with vertices $(0,.5)$, $(.5,.5)$, $(.5,1)$. This right triangle, with base and altitude both equal to .5, has area equal to $1/8$. So the probability that X and Y in case 1 are such that a triangle is possible, that is, that (X, Y) lies in the right triangle of area $1/8$, is $1/8$. Now we must consider case 2: $0 < Y < X < 1$. By symmetry, this case must yield the same answer, $1/8$. The only other case $0 < X = Y < 1$ may be ignored since this event is a straight line in the unit square and so has area (probability) zero. The answer to the problem is the sum of the probabilities of case 1 and case 2: $1/4$.

11.7   Buffon's needle problem

The Bertrand paradox already discussed is a problem combining geometric ideas with probability. The branch of probability theory dealing with such questions is not surprisingly called *geometric probability*. Perhaps the oldest problem in this field was the Buffon needle problem stated in 1777. It goes like this: a table, the floor, or part of any plane surface is ruled with parallel lines D units apart. A needle of length $L \leq D$ is tossed at random onto the surface. The needle can either intersect one of the ruled lines or lie in the strip between a pair of them. Find the probability of the needle intersecting a line.

To solve the problem, draw a picture of the needle lying on the surface (see Fig. 11.5). Let X be the distance of the midpoint of the needle to

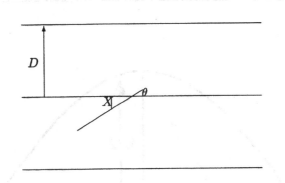

FIGURE 11.5. Needle of length $L \leq D$ lying across a line

the nearest of the ruled lines, and let θ be the angle between the positive direction of the ruled line (going off to the right) and the needle. The angle θ can be anything from 0 to 180 degrees. It is more convenient to measure angles in terms of radian measurement rather than degrees—the conversion factor is given by 180 degrees $= \pi$ radians. Therefore, by specifying X between 0 and $D/2$ and θ between 0 and π radians we have specified the direction of the needle. Tossing the needle at random is taken to imply the independence of X and θ, each with the uniform distribution. Draw a right triangle using the side of length X, and the ruled line as bases and the needle (extended if necessary) as the hypotenuse of length h. The needle intersects a ruled line if $h < L/2$. Elementary trigonometry shows

$$\frac{X}{\sin \theta} = h < \frac{L}{2},$$

that is, $X < (L/2) \sin \theta$. Consider a $\theta-X$ plane with θ as the horizontal axis and X as the vertical axis. If we plot in this plane the curve $X = (L/2) \sin \theta$, we get a picture that looks like Fig. 11.6. The event "the needle intersects a line" is equivalent to the event "X lies above the θ axis and below the curve $X = (L/2) \sin \theta$." As in the preceding problem, the uniform distribution and independence mean that the area below the curve relative to the area of the rectangle with base π and height $D/2$ is the desired probability. Using calculus, it is possible to show the area below the curve is equal to L (this simple answer is due to the use of radian measurement). The area of the rectangle is $\pi D/2$, so the probability of the needle crossing a line is $2L/(\pi D)$.

Now start to toss the needle at random onto the ruled surface in independent trials. Let U_i be indicator functions equal to 1 if the needle hits a line on trial i and 0 if the needle does not hit a line on trial i. The sum of the first n variables U_i gives the total number of the first n trials in which the needle intersects a line. In Chapter 8, we saw how these variables U_i

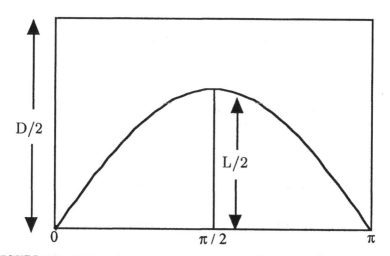

FIGURE 11.6. Area under curve represents probability of needle hitting a line

can give us a frequency interpretation of the probability by using the law of large numbers: with probability one, we have

$$\lim_{n \to \infty} \frac{\text{number of times the needle hits a line in } n \text{ tosses}}{n}$$

$$= \lim_{n \to \infty} \frac{U_1 + U_2 + \cdots + U_n}{n}$$

$$= EU_1 = P(\text{needle hits line on first toss}) = \frac{2L}{\pi D}.$$

This relation gives us a method for estimating π. Toss the needle many times, say n. Let m be the number of times the needle hits the line in the n tosses. Then from the above

$$\frac{m}{n} \approx \frac{2L}{\pi D},$$

which is equivalent to

$$\pi \approx \frac{2Ln}{Dm}.$$

This procedure may seem strange since we are estimating a non-random quantity π using a random device. This is, however, the basis of a very powerful and useful technique in modern mathematics, called the *Monte Carlo method*, whereby computations sometimes too difficult to be obtained by direct approaches are solved by means of computers and probabilistic reasoning. We'll return to this topic in Chapter 13.

11.8   Exercises for Chapter 11

1. Buses leave a bus stop at at 6, 6:15, 6:30, 6:45, and 7 pm. If Felix and

Alice each arrive randomly and independently at the bus stop during the hour from 6 to 7, what is the probability of their being on the same bus?

2. Suppose you break a stick at random at some point on it. What is the probability that the larger piece is more than twice as long as the shorter piece?

3. Solve the problem given in the last sentence of section 11.5. (Hint: Let the chord be determined by its midpoint chosen randomly inside the circle.)

4. Give solutions of Bertrand's paradox if the inscribed equilateral triangle is replaced with an inscribed square.

5. Suppose w is a normal number between 0 and 1. Take any integer $n > 0$ and consider the number $w_n = 10^n w$. Does the decimal part of w_n have to be normal? What can you say about w_n if w is not normal?

6. Find the probability of the following events that refer to the infinite decimals in the interval from 0 to 1: (a) The even digits are never equal to 0. (b) The 100th and 200th digits are both larger than 5. (c) At least one of the first two digits is equal to 0.

12

Normal Distributions, and Order from Diversity via the Central Limit Theorem

> Oh! Blessed rage for order, pale Ramon,
> The maker's rage to order words of the sea,
> Words of the fragrant portals, dimly-starred,
> And of ourselves and of our origins,
> In ghostlier demarcations, keener sounds.
>
> Wallace Stevens, *The Idea of Order at Key West*

12.1   Making sense of some data

As the head of a team of scientists, you are interested in compiling data on the heights of adult American men (by "American" we mean those who live within the borders of the United States). To do this, you would first want to choose a large sample of this population according to scientific principles ensuring that the sample is *random*. We will have more to say about randomness in the next several chapters, but essentially what you want is a sample representative of the entire population under study. So, for example, since you are interested in the population of adult American men your sample will not consist exclusively of men living in the Bronx in New York, nor will it consist solely of men residing on the west coast, or of those who earn more than $100,000 a year, or, in general, any category whatsoever restricting the sample from being representative of the entire population. One way you could go about choosing a random sample is to assign a unique number to each adult male in the population, and then select a certain portion of these numbers using a random device such that each number has the same probability of being selected. For example, you could write the numbers on cards, toss all the cards into a (very large) hat, mix well, and then select numbers with replacement (that is, after each selection you put the selected card back into the hat, mix again and select another card, etc.). Replacement of the cards is necessary to ensure that each card always has the same probability of being selected. If you happen to select a card that you had already chosen previously, it corresponds to an adult male chosen before, so we just ignore this observation, replace the card, mix, and keep repeating until a previously unchosen card is selected. In practice, of course, this procedure would be impossible to carry through—for one thing you would be hard pressed to find a hat big enough to hold the cards! It is also very hard to mix a lot of things so that a reasonable approximation to a uniform distribution model is achieved—see Section 13.6 for an interesting example. Statisticians have more sophisticated ways of finding a random sample; we'll take another look at this in Chapter 14. For now let's suppose you have this sample. What you are going to do is measure these individuals as accurately as you can and record these measurements. You will have a bunch of very large books in which each person's height is written down, and when the head of the Department of Demographics comes to your office and asks "What can you tell me about the height of the American adult male?" you will proudly haul out the books and say, "Take a look—it's all in there."

Unfortunately, these raw data are too cumbersome to handle. There is too much of it for it to make any sense. You will not be able to see the forest for the trees; no patterns or tendencies are apparent because of the welter of detail. So the head of the Department of Demographics will go home with a big headache and no insight. On his way out, he hoarsely pleads with you, "Can't you summarize all that stuff in a form a little easier to grasp?" After some thought, you get an idea. You take a horizontal x axis and label it starting at an integer A. Suppose you are measuring heights to the nearest .1 of an inch. Then each unit on your axis can represent .1 of an inch, and you mark off units from A until you get to another integer B. You can label A and B as the minimum and maximum values of the data. So now you have a line segment from A to B as your height axis—each point on the line segment represents a possible height in the data. What you want to do now is define so-called *class intervals*, that is, break down the height axis into a number of intervals. Suppose your data has a minimum height of 62 inches and a maximum height of 76 inches. We could define class intervals with a 2 inch length by specifying the first interval from 61.55 to 63.55, the second interval from 63.55 to 65.55, etc., with the last interval going from 75.55 to 77.55. The reason for picking the endpoints of the interval to two decimal places is so that each item of data, which is measured to only one decimal place, cannot fall on an interval endpoint, and unambiguously belongs to a unique interval. Now you and your staff go through the books of raw data and count the number of data items lying in each class interval.

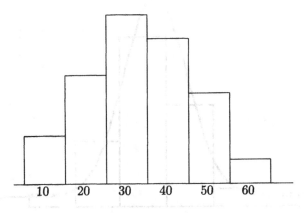

FIGURE 12.1. A typical histogram. The numbers represent the midpoints of class intervals

You then construct a bar over each interval with height proportional to the number of data items lying in that interval. The proportionality constant is a fixed number chosen to make the sum of the areas over all intervals equal to 1. The area of the bar over each interval is therefore equal to the proportion of individuals in the sample with height lying in that interval. Congratulations—you have just reinvented the *histogram* (see Fig. 12.1).

The above description is a very rough indication of the ideas involved in constructing one histogram from the given data. There are many other histograms you could have constructed; which one you want depends on your goals. The 2 inch length intervals give you eight class intervals. If you don't need that fine a breakdown, you could have used class intervals of length 4 inches, say, obtaining only four class intervals. In addition, before deciding on your interval range, you will generally want to study your data carefully and make some adjustments. For example, you may have one person who is 90 inches tall with all the other data items at or under 76 inches. It doesn't make much sense in extending the height axis all the way to 90 and getting a lot of empty class intervals. One way to handle this is simply to exclude the single extreme value 90 from the data on which the histogram is based (you lose information by excluding values, of course, so you should keep track of these extreme values for your final report—they may turn out to be important).

The histogram is a wonderful device for summarizing the data and presenting it in a form that shows important features. One look at the histogram and we can see which intervals contain relatively large numbers of data items and which contain relatively few. If the sample is large enough, the Law of Large Numbers says that the area of any bar over an interval,

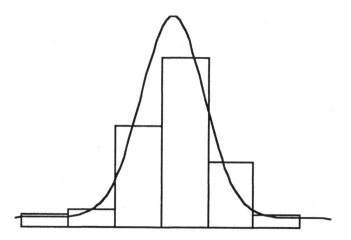

FIGURE 12.2. Bell-shaped curve approximating a histogram

namely, the proportion of the sample lying in the interval, is an approximation to the probability of a sample measurement falling into that interval. If the sample is representative of the entire population, the area of the bar should be an approximation to the probability that an American male chosen at random has height lying in that interval.

One thing we frequently want to do with histograms is approximate them by a continuous bell-shaped curve representing a normal distribution; we will start discussing these distributions in the next section. A crude example of this is seen in Fig. 12.2, where we just have a few class intervals with rather large length. If the sample is large enough, the number of intervals is taken large and the length of the intervals is sufficiently small, we can find a bell-shaped curve which does a good job approximating the histogram. What this means is that the area under the curve over any class intervals is a good approximation to the area under the bars for those intervals. The more intervals and the smaller their length, the better the bell-shaped curve approximates the histogram. Therefore, the area under the bell-shaped curve over an interval is an approximation to the probability of a measured height lying in that interval. From what we have learned about continuous distribution functions, this bell-shaped curve can be used as an estimated density function for the random variable $X =$ height of an adult American male.

12.2   The normal distributions

Now let us change the scene to the seventeenth and eighteenth centuries when Abraham DeMoivre and Pierre Laplace were working on various problems in probability and were led to a certain continuous distribution func-

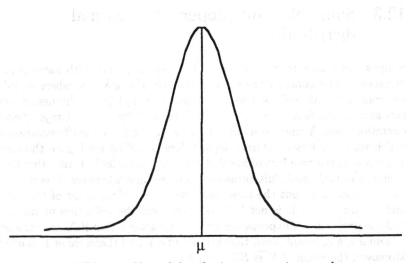

FIGURE 12.3. Normal distribution symmetric around μ

tion which we will call the *standard normal* distribution. This distribution belongs to a random variable which, for reasons of tradition, we will designate by the letter Z. The density of the standard normal distribution is a function whose exact mathematical form is not important for our purposes and would require some explanation, so we omit giving it. If this density is graphed in the x-y plane, a bell-shaped curve is obtained of the type described in the preceding section. In this case, it turns out that $EZ = 0$ and $\sigma^2(Z) = 1$, and the curve is symmetric around 0.

Upon further investigation, it was discovered that there was really a whole class of related distributions which could be defined by writing down a density function $f(x, \mu, \sigma)$ depending upon two parameters μ and σ. Here μ is any real number and σ is any positive number, and for each choice of a μ and a σ a particular distribution is determined. This class of related distributions was called normal, or Gaussian (after Karl Friedrich Gauss, who used it a lot). The class of normal distributions is called a two-parameter family of distributions because a particular member of the class is determined when the two parameters μ and σ are fixed.

If $\mu = 0$ and $\sigma = 1$, then we get the standard normal distribution. In general, it was observed that if X has the distribution $f(x, \mu, \sigma)$, then $EX = \mu$ and $\sigma^2(X) = \sigma^2$, and the distribution is symmetric around μ (see Fig. 12.3). Note that we use the notation $\sigma^2(X)$ for the variance of the random variable X and σ^2 alone for the numerical value of a variance.

12.3   Some pleasant properties of normal distributions

Suppose we take a random variable X which is normal with mean μ and variance σ^2 and consider the random variable $Y = aX + b$, where a and b are constants. We will assume $a > 0$ here to simplify the discussion, but this assumption is not necessary. Multiplying X by a is a *change-of-scale* operation—one X unit becomes a Y units. Adding b is a *shift* operation in which values of Y are obtained from values of aX by adding on the same amount b. It is a nice fact of life that the random variable Y resulting from a change of scale and shift turns out to be normal whenever X is normal. We can also figure out the mean and variance of Y in terms of the mean and variance of X. It is not hard to show from the definition of variance and the properties of expectation given in Chapter 8 that if X has variance σ^2 and a is a constant, then the variance of aX and therefore of Y is $a^2\sigma^2$. Moreover, the mean of Y is $EY = a\mu + b$.

To get an idea of what happens to the density of a standard normal variable Z when you perform a shift and a change of scale on it, first consider the shift $U = Z + b$ where b is constant. The density of U is identical in shape to the standard normal density; it has only been shifted so that its peak value is at b rather than at 0. The variance of U is the same as the variance of Z (the spread of values around the mean is intuitively the same since we just shifted one density over to get the other).

Now suppose we look at the change-of-scale operation $V = aZ$. Then the density of V is still centered at 0, but the variance of V is a^2. If $a > 1$, the values of V are more spread out around 0 than those of Z, but if $a < 1$ the values are less spread out around 0. It is apparent that the most general normal variable X with mean μ and variance σ^2 may be obtained by performing a shift and change of scale to Z, namely, put $X = \sigma Z + \mu$.

Figure 12.4 shows three densities. The first is centered at 0. The second is a shift of the first to a new center μ_1 keeping the variance, hence the shape, the same. The third involves both a change of center, that is, a shift, as well as a change of variance. In this case, the variance of the third density (c) is smaller than the variance of (a), so that (c) is more concentrated around its mean.

Conversely, if X is given to be normal we can transform it to the standard normal Z by a shift and change-of-scale transformation. If X has parameters μ and σ, the random variable

$$(X - \mu)/\sigma = (1/\sigma)X - \mu/\sigma$$

is obtained from a shift and change of scale applied to X, and has mean 0 and variance 1, so must be standard normal. This is an extremely useful fact since it allows us to rephrase any question about an arbitrary normal distribution in terms of an equivalent question about the standard normal distribution.

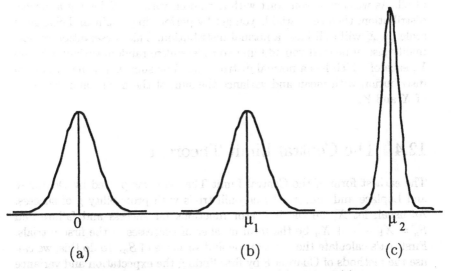

FIGURE 12.4. (a) Normal density centered at 0. (b) Shift of (a) to new center μ_1. (c) Shift of (a) to new center μ_2 with decrease in variance

To pin matters down, let's say we want to calculate a probability statement about the normal variable X with parameters μ and σ^2; for example, suppose we want to get $P(a < X < b)$. Then

$$P(a < X < b) = P\left(\frac{a - \mu}{\sigma} < Z < \frac{b - \mu}{\sigma}\right),$$

using a little algebra and the relation $Z = (X - \mu)/\sigma$. The right-hand side is a probability involving a standard normal variable. For this reason, it is sufficient to have tables compiled for the standard normal distribution. This often takes the form of listing at intervals of .01 the numbers $a = .01, .02, \cdots$ up to about 3 or 4, and with each of these the area under the density over the interval from 0 to a, say, or from $-\infty$ to a. From this information and by making use of symmetry, you can calculate the approximate probability of the standard normal variable Z lying in any interval. In particular, such a table shows that about 68 percent of the area lies within one standard deviation of 0 and about 95 percent lies within two standard deviations; that is, the standard normal variable Z satisfies

$$P(-1 < Z < 1) \approx .68, \quad P(-2 < Z < 2) \approx .95.$$

From this, it is not hard to check that for *any* normal variable about 68 percent of the area lies within one standard deviation of the expectation and about 95 percent lies within two standard deviations.

Since this section was about pleasant properties of the normal distribution, we should not end it without mentioning one of the most pleasant of all. As we saw, if you start with a random variable X having a normal distribution, then the variable you get by performing a shift and change of scale on X will still have a normal distribution. This preservation of normality also appears if you add up two *independent* random variables X and Y, each of which has a normal distribution. The sum $X + Y$ has a normal distribution with mean and variance the sum of the means and variances of X and Y.

12.4 The Central Limit Theorem

The earliest form of the Central Limit Theorem was proved by DeMoivre and Laplace and concerned Bernoulli trials with probability p of success. As usual, let X_i be the indicator functions for success and failure and $S_n = X_1 + \cdots + X_n$ be the total number of successes in the first n trials. First, let's calculate the expectation and variance of S_n. To do this, we can use the methods of Chapter 8 by first finding the expectation and variance of X_1. $EX_1 = p$, and

$$\sigma^2(X_1) = E(X_1 - EX_1)^2 = E(X_1 - p)^2 = p(1 - p).$$

Since the expectation of S_n is the sum of the expectations of the X's and the variance of S_n is the sum of the variances of the X's, we have shown:

> The total number of successes S_n in n Bernoulli trials with success probability p has expectation np and variance $np(1-p)$.

The idea now is to *standardize* S_n by forming the random variable

$$Q_n = \frac{S_n - np}{\sqrt{np(1 - p)}}. \tag{12.1}$$

The random variables Q_n each have mean 0 and variance 1 (whenever we have any random variable with finite mean and variance, if we subtract the mean of the variable and divide by the standard deviation we always get a standardized variable with mean 0 and variance 1). DeMoivre and Laplace proved that the distributions of the variables Q_n converge to (that is, get closer and closer to) the standard normal distribution. We can write this in the following way: for any interval from a to b

$$\lim_{n \to \infty} P\left(a < Q_n < b\right) = \tag{12.2}$$

area under the standard normal density over the interval from a to b.

What is all this saying exactly in plain English? It says that if you take the indicator functions X_i for Bernoulli trials and add enough of them up, and then standardize the sum to have mean 0 and variance 1, then the random variable you get will be almost standard normal. So the standard normal distribution appears somewhat mysteriously as a limiting form of a binomially distributed random variable that we standardized. This is an amazing and beautiful result. On a theoretical level it seems to open up something new and very exciting. How about sums of independent random variables not necessarily connected to a Bernoulli trial set-up—do these sums, when properly standardized, approximate a standard normal or perhaps some other distribution? It took about 150 years before an astounding answer to that question was fully proved in a wide generalization of the DeMoivre-Laplace theorem. Here is how it goes: assume $S_n = X_1 + X_2 + \cdots + X_n$ is the sum of independent, identically distributed random variables with common expectation μ and common variance σ^2. So the expectation of S_n is $n\mu$, and the variance of S_n is $n\sigma^2$. This time, let us form the analog to formula 12.1 in the general case: instead of np we have $n\mu$ and instead of $\sqrt{np(1-p)}$ we have $\sigma\sqrt{n}$, to get

$$Q_n = \frac{S_n - n\mu}{\sigma\sqrt{n}}. \tag{12.3}$$

The Central Limit Theorem asserts that the relation of formula 12.2 still is true.

Why is this such a startling result? Because we can start out with independent random variables X having *any* distribution whatsoever, as long as there is a finite mean and variance. The theorem says that if we add up enough of these and standardize, we are going to get an approximately standard normal variable. So in spite of the initial arbitrariness of the X's, order emerges after awhile, with the standard normal distribution playing a curiously special role as a limiting distribution. The DeMoivre-Laplace theorem relating to Bernoulli trials is a very special case of the general version of the Central Limit Theorem just given.

It is worthwhile to look at what the Central Limit Theorem says about the sums S_n of the independent, identically distributed random variables X_i with finite mean and variance. Since Q_n has a distribution close to a standard normal variable Z for large n, then, according to the discussion in the preceding section, we might expect the change of scale and shift operation

$$S_n = \sigma\sqrt{n}Q_n + n\mu$$

to have an approximately normal distribution with mean $n\mu$ and variance $n\sigma^2$. But it will be hard to say anything meaningful about these non-standardized sums because the variance is getting large without bound, and if $\mu \neq 0$ the means are moving to plus or minus infinity. If, however, the distributions of the X's are not always fixed but change as n increases in such a way that the variance of S_n converges to a finite value, then we may get something useful about the normality of S_n for large n. We will see this happen in the discussion of Brownian motion in Chapter 17.

The Law of Large Numbers and the Central Limit Theorem are undoubtedly the two most important theoretical results of the theory of probability. The Central Limit Theorem is called a *weak* limit theorem; that's the probabilist's way of saying that it is a statement about the convergence of distributions to a distribution. A *strong* limit theorem is about the convergence of averages—for example, S_n/n— for particular sample paths or realizations in actual plays of a game of chance. In our discussion of the Law of Large Numbers, it was the Strong Law that was of interest to us in talking about gambling. But in trying to give a plausibility argument for the Strong Law, we just gave an argument for the Weak Law in the Appendix to Chapter 8. The weak version of the Law of Large Numbers can be expressed in the following way: the distributions of the variables

$$W_n = \frac{X - n\mu}{n} \tag{12.4}$$

converge to the distribution of the random variable which is identically 0 (we say such a variable is *degenerate* at 0 because the variable only takes on one value with probability 1). Let's compare this to the Central Limit Theorem. This asserts the convergence of the distributions of

$$Q_n = \frac{X - n\mu}{\sigma\sqrt{n}} \tag{12.5}$$

to the standard normal distribution. Notice that the numerators (the top part) of the fractions in formulas 12.4 and 12.5 are identical. In formula 12.5, we see that as n increases the numerator is increasing in such a way that the factor $\sqrt{n}$ essentially standardizes the variables (the σ can be ignored here) so that the normal distribution is a limiting distribution. On the other hand, if we divide by n rather than $\sqrt{n}$ as in formula 12.4, we are dividing by a factor so large that the quotient random variables look more and more like the random variable degenerate at 0. It will follow from the Central Limit Theorem that dividing by the factor n^r rather than $\sqrt{n} = n^{\frac{1}{2}}$ in formula 12.5 gives variables looking more and more like the random variable degenerate at 0 whenever $r > 1/2$ and gives variables looking more and more like the random variable degenerate at either plus or minus infinity whenever $r < 1/2$ (in other words, the variables are getting large without bound in absolute value). Therefore, 1/2 turns out to be the right or critical exponent for n, so that when we divide we get variables looking more and more like a non-degenerate distribution, namely, the standard normal.

12.5   How many heads did you get?

To illustrate the practical application of the ideas in this chapter, consider the following problem.

The prisoner of Chapter 3 is bored, so he tosses a fair coin 1000 times. What is the probability of at least 495, but at most 510, heads showing up?

The binomial distribution tells us how to get the probability of getting exactly i successes in 1000 tosses of a fair coin (probability of success = .5), so the problem poses no theoretical difficulty; simply calculate this probability for each value of i between 495 and 510 and then add these probabilities up. It is the practical matter of calculation that causes the unpleasantness here, however. You must calculate nasty things with lots of factorials like $C_{1000,i}$ for i between 495 and 510. There are ways to approximate these factorials, but then you would not get the exact answer, only an approximation, and you would be doing a lot of work. If you are willing to settle for an approximate answer, here is a relatively easy way to get one using the Central Limit Theorem.

Let X_i be the indicators, with values 1 or 0 depending on whether the ith toss yielded head (success) or tail (failure), respectively, and use the DeMoivre-Laplace version of the theorem. We have $p = P(X_1 = 1) = P(X_1 = 0) = 1 - p = .5$, so from Section 12.4 we know that the random variable

$$S_{1000} = X_1 + X_2 + \cdots + X_{1000}$$

has expectation $1000 \cdot (1/2) = 500$ and variance $1000 \cdot (1/2) \cdot (1/2) = 250$. Therefore

$$Q_{1000} = \frac{S_{1000} - 500}{\sqrt{250}}$$

should be approximately standard normal since 1000 is a reasonably large value. The problem requires that $495 \leq S_{1000} \leq 510$, and this implies

$$\frac{495 - 500}{\sqrt{250}} \leq Q_{1000} \leq \frac{510 - 500}{\sqrt{250}}.$$

Use a calculator to find that this relation is approximately the same as $-.32 \leq Q_{1000} \leq .63$, and then the Central Limit Theorem gives

$$P(-.32 \leq Q_{1000} \leq .63) \approx P(-.32 \leq Z \leq .63),$$

where Z is the standard normal variable. At this point, we turn to a table of the standard normal distribution to find $P(-.32 \leq Z \leq .63) \approx .3612$. Therefore, the probability that Q_{1000} lies between $-.32$ and .63, or equivalently, that S_{1000} lies between 495 and 510, is approximately .3612.

12.6   Why so many quantities may be approximately normal

Many quantities of interest in the real world that we measure have approximately normal distributions. Among these, we find the heights, weights, and blood pressures of a population of people; test scores; the length of life of certain electrical or mechanical components; etc. Why does the normal distribution turn up so frequently as an empirical fact of life?

The Central Limit Theorem is sometimes used to give a theoretical explanation for the frequency with which normal or approximately normal distributions describe natural phenomena. It is said that the height of an adult, for example, is due to a multitude of causes: genetic makeup, diet, environmental factors, etc. These factors often combine in an approximately additive way so that the result is, by the Central Limit Theorem, close to normally distributed. It is true that all these factors contributing to an individual's height do not in general have the same distribution nor are they always independent, so the version of the Central Limit Theorem discussed here may not apply. There are, however, generalizations of the Central Limit Theorem valid when there are departures from the identically distributed assumption, and even from the independence assumption. Such results could offer a reasonable explanation of why many phenomena are approximately normally distributed.

There are a few other points we should mention in ending this chapter. First, whenever we have a limit theorem telling us that some sequence of numbers gets close to some number, it is always important to have an idea of how far out in the sequence you have to go before expecting the sequence to be close to the limiting number within some desired degree of accuracy. This problem was mentioned in Section 8.5 with reference to the Law of Large Numbers. Similarly, the approximation to the standard normal distribution by the distributions of the variables Q_n given by the Central Limit Theorem would be of little practical value unless we had an estimate for the error between $P(a < Q_n < b)$ and $P(a < Z < b)$ for given n. Such estimates exist; the size of the error for any given n depends upon the common distribution of the X's. As with the Law of Large Numbers we omit the technical details. In the preceding section, we tacitly assumed the number of tosses, 1000, was large enough to give us a reasonably accurate estimate.

Second, if you know you have something, like the heights of Section 12.1, with a normal distribution, how do you go about finding out which one of the infinite possible normal distributions it is? The answer is that you will have to estimate the determining parameters μ and σ^2 from the data. This is a basic problem in statistical inference which we will be speaking about at length in Chapter 15, but to look ahead for a moment, suppose X_i is the height of the ith individual of the sample. Because the sample was random, the sequence $X_1, X_2, \cdots$ is an independent sequence of random variables, and statistical criteria show the *sample mean*

$$\overline{X} = \frac{X_1 + X_2 + \cdots + X_n}{n}$$

is a good estimator of the unknown population expectation μ and the *sample variance*

$$s^2 = \frac{(X_1 - \overline{X})^2 + (X_2 - \overline{X})^2 + \cdots + (X_n - \overline{X})^2}{n - 1}$$

is a good estimator of the unknown population variance σ^2. These estimators become better as n increases.

12.7   Exercises for Chapter 12

1. A random variable X has the normal distribution with $\mu = 67$ and $\sigma = 2$. Find $P(130 < 2X < 136)$ in terms of the standard normal variable Z.

2. A fair coin is tossed in repeated independent trials. Let S_n be the total number of heads after n trials. Standardize S_n and use this to solve the following problem: consider the ratio $S_n/\sqrt{n} = R_n$ and let x be any fixed number. Estimate $P(R_n < x)$ in terms of a standard normal variable Z if the number of trials n is sufficiently large.

3. Consider a gambler who wins or loses \$1 with probabilities p and q, respectively. Assume the gambler is allowed to keep playing (and does so) even if he goes into debt. Let S_n be the gambler's accumulated winnings after n games. (a) Find the expectation and variance of S_n. (b) Standardize S_n and use the Central Limit Theorem to say something about this standardization.

4. Use part (b) of exercise 3 to show that, given any fixed number $1 - \varepsilon$ as close to 1 as you please ($\varepsilon > 0$ and as small as you want), there exists some constant $K > 0$ that depends on ε such that for all n sufficiently large we have

$$P(S_n < n(p - q) + \sqrt{n} \cdot K) > 1 - \epsilon.$$

5. Use exercise 4 to show that a gambler playing an unfavorable game who is allowed to amass debt will, as he continues playing, sink deeper and deeper into debt with overwhelming probability. (Hint: a number of the form $an + b\sqrt{n}$ where $a < 0, b > 0$ will, for large enough n, be less than any fixed negative number.)

6. Let Z be a standard normal random variable. Describe the distribution of the random variable $U = -Z$.

13

Random Numbers: What They Are and How to Use Them

> The lot causeth disputes to cease, and it decideth between the mighty.
>
> *Proverbs* 18:18

13.1   What are random numbers?

In 1955, the Rand Corporation published a book listing a million random digits (see [27]). A typical page contains hundreds of the digits 0 through 9 written for easy reference in little matrices (squares) having five digits along each row and each column. We are interested in two main questions: what are random numbers and why do we need them?

The most basic way to define random numbers is that they are numbers generated by a random procedure involving repeated independent trials. When we speak about the random digits 0 through 9 it is assumed that a trial of the procedure yields each of the ten digits with probability .1. Here is how we might generate these random digits. Suppose we take ten cards of the same size and write the digits 0 through 9 on the cards so that each card has a different digit. Then take a large hat, say, toss in the cards, and mix well. Now choose a card at random from the hat, that is, reach in without looking and choose. Write down on a piece of paper the digit appearing on the card you have chosen. Put the card back into the hat and mix the cards again. Repeat the procedure by choosing a card at random, writing down the digit appearing on the card, replacing, mixing, choosing again, and so on. The string of digits you are writing down constitutes a string of random digits because it has been produced by a random device supposed to yield each digit with probability .1 in independent trials (the procedure might not really give digits with equal probability because of poor mixing or other reasons—see, e.g., Section 13.6). You could have produced the random digits by an equivalent method by using, for instance, a modified roulette wheel in which the wheel has been divided into ten equal parts, each one corresponding to one of the ten digits. The book of random digits published by the Rand Corporation was produced by a more sophisticated version of this roulette wheel, one in which electronic pulses are used instead of spinning disks.

Given random digits, how do we get more complicated random numbers? Suppose you had generated the sequence 3217900597 by reading ten digits from the table. Then each digit is random, and furthermore the two-digit numbers 32, 17, 90, 05, 97 obtained by taking the numbers two at a time are random numbers because they have been produced by a random procedure ensuring that the one hundred two-digit numbers 00 through 99 each have probability .01 of appearing and, moreover, the selections of these two-digit numbers are independent. This follows from the way we selected the individual digits, namely, in a uniform, independent manner. Thus, taking the original ten-digit sequence and choosing the numbers two at a time going backward to get 79, 50, 09, 71, 23 also give random two-digit numbers, as does any other way you might think of to get two-digit numbers using the generated string, as long as the method does not use the same selection more than once. If you need five two-digit random numbers you can use any of the methods to generate them. None of these methods is any better than any other, so you may as well take the simplest one: 32, 17, 90, 05, 97, reading from left to right.

Note that if you need ten random two-digit numbers, which requires twenty digits all together, you should select 20 random digits from the table, not use, for example, 32, 17, 90, 05, 97, 79, 50, 09, 71, 23 by selecting digits read both forward and backward from the same initial ten-digit string. Randomness is lost when the same digits are used more than once in this way. This is evident above because the first five two-digit numbers give us information about the whole sequence, (for instance, that 6 does not occur), and so following selections are neither equally likely nor independent.

For the same reason, we would not want to use the same lines or the same page of the random number table over and over again every time we need random numbers. If we do this, the unpredictability associated with randomness is lost. With the Rand table you have a million digits on many pages. You should use this abundance to strive for unpredictability, that is, randomness. One way you could select digits from the Rand book is by starting at the first page and reading line by line (or column by column) until you have enough digits, then marking where you stop. The next time you need digits, begin reading where you left off the first time, and so on. In this way, you will go through the book. Another method starts by using a random number table to select for you a random page and a random line number where you will start reading the digits you will actually use for your purpose. The basic idea in all this is that you want to approximate independent, identically distributed selections as closely as possible, and this goal will be compromised if you use the same limited set of digits all the time.

We have seen that a random number is just a number produced by a random procedure. Another way to say this is that a random number is the value of a random variable X having some distribution. In the most basic case of random digits, $X = 0, 1, \cdots, 9$ has the discrete uniform distribution putting probability .1 at each digit. A list of such random digits is generated by repeated observations $X_1, X_2, \cdots$, of independent random variables all with the same distribution as X. From these lists, such as those in the Rand book, more complicated random numbers can be constructed by various mathematical techniques. An example of this appears above where $X = 00, 01, \cdots, 99$ with the uniform distribution. If you wanted $X = 1, 2, \cdots, 100$ with the uniform distribution, you could identify 100 with 00, 1 with 01, 2 with 02, etc., up to 99 with 99. In this way, a selection of two-digit random numbers corresponds to a selection of numbers from 1 to 100.

But now suppose we want to do something a little harder, like picking random numbers between 1 and 12. Using random digits we can accomplish this in the following way. First, form two-digit random numbers using a list of random digits while crossing out all pairs not equal to one of the values 01, 02, $\cdots$, 12. As an example, consider a table of random numbers obtained from the Rand Corporation book:

$$
\begin{array}{ccccc}
12 & 66 & 00 & 95 & 71 \\
81 & 85 & 26 & 06 & 20 \\
03 & 06 & 86 & 13 & 17 \\
29 & 62 & 35 & 85 & 30 \\
30 & 52 & 05 & 05 & 88 \\
\end{array}
$$

If we cross out all pairs corresponding to numbers bigger than 12, the pairs 12, 00, 06, 03, 05 are left after elimination, and I claim that these represent five random numbers taken from the uniform distribution with 12 outcomes. To see the truth of this assertion, observe that each pair has probability .01 of being selected from the sample space of the 100 two-digit numbers 00 through 99. Therefore, by conditional probability

$$P(12 \text{ appears/given only } 01, 02 \cdots, 12 \text{ can appear}) = \frac{.01}{.12} = \frac{1}{12},$$

(by crossing out all two-digit pairs not specified by the conditioning event, the reduced, or conditional, probability space is obtained). In the same way, the conditional probability of each of the pairs 01 through 12 can be seen to be 1/12. Moreover, the randomness of the original digits implies the independence of the selected pairs. The table above has given us five random numbers of the type we desire. If more than five are required, choose as many matrices of random digits as you need so that by repetition of the method you have as many random numbers of the desired form required. We can, in a similar fashion, generate random numbers from any finite uniform distribution using a table of random digits.

Random digit tables can also give us values, to any degree of approximation, of a random variable X having a continuous distribution. Let's look at this for a uniform distribution on the unit interval 0 to 1. Let X be a number chosen from the unit interval in accordance with the continuous uniform probability distribution, and recall (from Chapter 12) that the digits of the decimal expansion of X are independent random variables with the finite uniform distribution, putting .1 at each of the ten possible values. Conversely, given that the digits of the decimal expansion of X are chosen independently according to the finite uniform distribution, it can be shown that X is a random variable with the continuous uniform distribution on the unit interval. This means that the number X chosen according to a uniform distribution can be approximated by taking a finite sequence of random digits and sticking a decimal point in front. For example, suppose we wanted to choose five numbers in the unit interval chosen uniformly and independently in the unit interval. If we are willing to settle for ten-digit approximations of each number, we can use the little table given above to construct the five numbers .1266009571, .8185260620, .0306861317, .2962358530, and .3052050588 by choosing each of the five rows of the table. If we are willing to stop after four digits, we can take (as one possible choice) .1266, .0095, .7181, .8526, and .0620 as approximations by taking digits along the top row until we run out, and then proceeding to the second row.

The Rand Corporation book also gives 100,000 "normal deviates," which are just values of a random variable X with a standard normal distribution. These were constructed from the random digits by explicit mathematical transformations. For any continuous distribution, values of a random variable having this distribution can always be obtained to any degree of approximation by performing mathematical transformations on random digits. The random digit tables, then, really give us all we need in the way of randomness to generate values from any distribution. Once we have the digits, the desired values depend on deterministic, not random, mathematical computations. These are easy to describe in theory but may be unpleasant to carry out in practice.

13.2   When are digits random? Statistical randomness

We have seen how to generate strings of digits that might reasonably be called random. Very often we are faced with the tricky converse question: given a string of digits, is it random? This is a question having no precise answer if we are simply given digits without any information about how they were generated, since the definition of random digits depends upon the procedure producing them. On the other hand, if we are given a string of digits, there should be some way to decide rationally whether the string conforms to our notion of what a random string ought to look like. What we need are statistical tests for randomness.

Let's tie all this down with a few dramatic examples. Suppose I present you with 10,000 digits all of which are equal to 0. Most likely, you would have grave doubts that they were generated by a random procedure where each digit has probability .1 of appearing. Your gut intuition finds the pattern of all digits 0 inconsistent with the notion of randomness because none of the digits other than 0 turn up in so many trials. Your guess as to the probability of randomness in this case (from a subjective probability standpoint) would likely be close to 0. If the 10,000 digits consisted of the pattern 0, 1, 2, 3, 4, 5, 6, 7, 8, 9 repeated in that order over and over until 10,000 digits were obtained, you would still judge the resulting sequence most likely non-random even though each digit occurs the right proportion of times. The trouble here is predictability and pattern, suggesting non-independence in the generating procedure.

Most sequences are not as extreme as the two examples just given. For most sequences, your intuition is not going to be particularly helpful in determining randomness. An average person not familiar with what random sequences actually look like is usually quite surprised that they frequently yield so many adjacent repeated digits. But adjacent repeated digits are not unusual in a random sequence since the event "one of 00, 11,···, 99 appears" has probability .1. And if we are examining a very long string of digits we should not be surprised if occasionally we find three or four adjacent repetitions, because as we know from Chapter 5, rare events occur from time to time and have high probability of occurrence if there are enough trials. Naively intuitive notions of randomness might find such repetitions suspect.

So we need impartial statistical tests to make decisions for us. In this case, it boils down to determining whether the observed sequence is so different from what we expect a random sequence to look like that we should reject it as random. With 10,000 digits, the expected number of occurrences of each digit under the assumption of randomness is $(.1)(10,000)=1,000$ times. The chi-square test is a statistical procedure that rejects a given sequence as random if its actual number of expected occurrences deviates significantly from the expected number under the assumption of randomness. This test would certainly reject the sequence of 10,000 zeros as random, and could be applied to any sequence. The second example above of the repeated pattern of the digits 0 through 9 has the right number of expected occurrences but fails to have other properties expected from random sequences. Here we could use statistical tests for *serial correlation*, a measure of the relationship between the digits of the sequence. In a random sequence, we would not expect 1 always to be followed by 2, for example. We could also use tests for *runs*, a run of length n meaning in this case that a particular digit occurs exactly n times in succession. As we have observed, a random sequence should have runs of length 2 fairly often, and the second example could be rejected as random because it has no such runs.

Statistical tests and statistical decisions are not, by their nature, exact enterprises, as we will see in Chapter 15. Many of them are based on the simple idea that, if what you observed has very small probability under a particular assumption, then you should reject that assumption. A statistical test can incorrectly lead to rejection of a hypothesis due to the observation of a rare event. There are thus a number of statistical tests that we might want a sequence to pass before we are willing to call it random. The usual procedure is to pick some of these which together check for all the basic properties we feel a truly random sequence should possess. We could use several different tests to check the same property, thereby getting added confidence in the randomness of the sequence. The sequence has to run the gauntlet of these tests and will be called random if it passes all or most of them. If the sequence fails one important test, it is rejected as random.

Even if a sequence of digits has been generated by what we believe is an appropriate random procedure, the sequence should still be subjected to statistical tests for randomness. This is necessary because a procedure could have built in biases even though it may, on the surface, appear to give values according to an equally likely distribution. Often adjustments have to be made to the original procedure, or mathematical transformations have to be applied to the digits, to get them to pass the various statistical tests (this was the case with the digits in the Rand book). In other words, real-life random procedures like picking cards from a hat, rolling a pair of dice, or spinning a roulette wheel are clearly only approximations to the mathematical idea of randomness, and sometimes adjustments have to be made to get real life more in line with the abstract model.

The preceding discussion suggests a more sophisticated notion of randomness. A sequence of digits could be called *statistically random* if it passes a battery of statistical tests for randomness. This kind of randomness ignores how the digits are actually generated; it only requires sequences of them to pass the statistical tests. The concept of statistical randomness is of the greatest importance for computers. It allows computer-generated *deterministic* sequences to substitute for actual *random* sequences because the deterministic sequences are statistically random. Statistical randomness is a much more useful idea than just plain randomness—it substitutes the extremely appealing and practical guide of statistical testing for the usually impossible task of determining how the digits were generated. It is operational in nature: if it *acts* random then it might as well *be* random. Let's now turn to this idea as realized on the computer.

13.3   Pseudo-random numbers

For a variety of tasks (some of which will be discussed in Section 13.5) it is very important to have access to a large table of random numbers when working on a computer. Clearly, it is desirable to have the computer itself generate these numbers in some manner. There are, however, certain problems associated with a truly random number generator built into the computer. One of the major ones is the following: if you are doing a job requiring a large bunch of random numbers and you want to reproduce your computations later, it will be necessary to *store* the random numbers. Finding the space for this is generally impossible, and even if it could be done it would use up huge amounts of memory. This major hurdle was cleared by a brilliant solution. Instead of generating actual random numbers using a random device, computers generate numbers that arise by iteration of certain special functions f. It works something like this: you take an initial "seed" value z_1 and the function f, the (so-called) random number generator, and let

$$z_2 = f(z_1), z_3 = f(z_2), \cdots, z_n = f(z_{n-1}),$$

thereby generating $\{z_i, 1 \le i \le n\}$, a deterministic (i.e., non-random) sequence, where n can be a large value. The function f is chosen so that the sequence $\{z_i\}$ turns out to be statistically random, so it acts just like a random sequence does in the sense that it passes certain statistical tests for randomness. Different seed values produce different statistically random sequences. You don't have to store anything (except the seed value) because the computer recalculates the same sequence if given the same seed value.

The numbers generated by f are called *pseudo-random* numbers. It is rather amazing that such functions f can be found. The sequence generated is as far from the original notion of random as possible—each term is completely determined given the preceding term. Yet the sequence is statistically random. The numbers generated by f are usually called random numbers even though they are not really random. For most scientific purposes, they provide a useful substitute for the real thing. But there are also problems with existing random number generators and there is a lot of research on finding good, fast ones. Accepting a pseudo-random sequence as statistically random only means that the sequence has passed a certain number of tests we have decided are important. Some well-accepted pseudo-random number generators have recently been observed to have subtle correlations and non-random patterns not detected by the statistical tests they passed (see [2]). This has led mathematicians to consider whether there might not be other, perhaps truly random, ways to generate random numbers quickly and efficiently so that the problem of reproducing results without storage could somehow be overcome. As of this writing, there has been little progress on this question.

13.4   Random sequences arising from decimal expansions

We already know (Chapter 11) that a uniformly distributed random variable X on the unit interval has a decimal expansion for which the digits are independent and, as we saw from the discussion of normal numbers, it is a consequence of the Law of Large Numbers that each digit appears with (limiting) frequency .1 for almost all values x of X. It is possible to conclude from this that the decimal expansion of almost all x in the unit interval produces a table of random numbers. In this case, we can say that the decimal expansion of x is random.

This conclusion may be theoretically very pretty but is useless as a way to generate random digits. As with normal numbers, if you use the uniform distribution to choose a point at random, you will almost certainly pick a "good" x, that is, one for which the decimal expansion is random, but for a given x there is usually no way to know whether it is good or whether it is one of the exceptions (remember, there are loads of exceptions—all the repeating and finite decimals, for instance). To make matters worse (as if they could be made any worse), there is the problem of how to choose the value of x uniformly from the unit interval in the first place. The way we usually do that was described earlier, by first getting random digits and stringing them together to get approximations. So we have a useless circle: getting random digits requires choosing uniformly from the unit interval, and to do that you need to have random digits.

At the moment, it is beyond mathematics to show that, for example, π has a random decimal expansion (or even that it is normal, as observed in Chapter 11—the normality of a number as we have defined it is a less stringent requirement than having a random decimal expansion). What we can do, however, is test a long, finite piece of the decimal expansion of π for statistical randomness. With π, we have a number whose decimal expansion has been carried out to more terms than any other number, and as far as I am aware the sequence of the digits appears to be statistically random. So although we cannot (at present) prove the decimal expansion of π random, we can get a certain degree of confidence that this is so from the statistical analysis of a long, finite piece of this expansion.

13.5   The use of random numbers

Now that we have random numbers, it is time to turn to the question of why we need them. One of the most important ways randomness has been used throughout history has been to choose people or objects in a way supposed to be free of bias or prejudice. The choice of a lottery ticket is made using a random device that determines the number of the winning ticket. From a pool of possible jurors, the selection of a jury is accomplished by picking names randomly from a physical device like a drum or by using a random selection obtained by computer (in this case, we are technically choosing a random name rather than number). Section 13.6 describes an interesting application of random techniques to choose soldiers in wartime. In statistical design and practice, randomness plays a critical role. One important way in which it is used is in the theory of random sampling of populations (in polling, for example). Here is a typical use of random numbers to select a random sample in the design of a scientific experiment.

Suppose we consider a biologist who is planning an experiment to test the effects of a tranquilizer drug. She has 100 laboratory mice and wants to divide them into two groups, 50 for the drug and the remaining 50 to act as a control group which will be given a placebo (sugar pill). The purpose of the experiment is to measure the effects of the drug, and only the drug, on the mice (by measuring a certain physiological property such as heart rate, say); therefore, the experimenter must try to exclude any factors other than the drug having a possible influence on her measurements. The mice getting the treatment (that is, the drug) should be as similar as possible to the mice receiving the placebo. If that is the case, then any difference between the two groups can more reasonably be ascribed to the treatment than to an extraneous difference between the groups. It would be wrong, for example, to take the 50 youngest mice and give them the treatment and use the 50 oldest for the control group since age could be a factor affecting heart rate, and any difference we measure between the groups may have more to do with this difference in age than with the administration of the drug. So we come to the question: how should the biologist choose the treatment group to eliminate as much as possible the chance for bias? The answer is that she should randomly select the mice going into the treatment group; this method of choosing is the best way of assuring the homogeneity of the two groups. Random selection is very easy using a table of random digits. In this case, first label the mice 00, 01, $\cdots$, up to 99. Then take the table and choose two-digit numbers until you get 50 distinct numbers (throw away any repeated numbers). The numbers you have chosen are the numbers corresponding to the mice the biologist should select for her treatment group—the mice not selected form the control group. Because the mice have been chosen at random, each mouse is just as likely to be in the treatment group as the control group, and any tendencies toward bias are minimized.

Someone may question whether such care to ensure a random selection is really necessary. He may argue that the biologist could pretty much get a random sample by just reaching into the cage and hauling out the first 50 mice in her reach. A little thought should convince this skeptic that any such selection procedure could easily introduce bias—the mice easily caught may be the more placid mice, those with a calm manner and a low heart rate. So while superficially it may seem that you are choosing randomly, unless you use a truly random scheme to select subjects, other methods can have subtle and dangerous pitfalls leading to bias in the experiment.

Another major area where random numbers are essential is in *simulation*. Suppose we want to toss a fair coin 1000 times and record the data. Instead of actually tossing the coin, we can let the computer simulate the experiment as follows. The computer has a random number generator which we can use to have it choose the digits 0 and 1 with probability .5 (we will see an algorithm, that is, a recipe, for doing this in the next chapter). Each random choice of a digit by the computer can be thought of as the toss of a fair coin, where 0 stands for head, say, and 1 stands for tail. We can then program the computer to make the random choice 1000 times and do whatever we want to do with the data, for example, count the frequency of the number of 0's appearing (corresponding to the number of heads). The data obtained should be statistically indistinguishable from actual data obtained from tossing a fair coin because the pseudo-random numbers from the computer are statistically random.

Using the computer's random number generator, programs can be written to simulate any process as long as the distribution involved is known. To simulate repeated rolls of a pair of dice, set up the generator to choose the integers 1 to 6 uniformly, and then generate these by pairs to get a roll. Roulette is easily simulated by generating the numbers 1 to 38 uniformly, and letting, for example, 37 stand for 0 and 38 for 00. Simulation of tossing the needle in the Buffon needle problem of Chapter 11 is not much harder. The random number generator has to be set up to give a value X uniformly distributed between 0 and $D/2$, where D is the distance between the parallel lines. In addition, we need a value of θ uniformly distributed between 0 and π radians. Whenever $X < (L/2)\sin\theta$, where L is the length of the needle, the selection of the pair (X, θ) represents a toss of the needle intersecting a line. Algorithms performing each of these simulations will be given in the next chapter.

Simulation is now an important part of many disciplines. Rather than actually building complicated devices (weapons, engines, aircraft, etc.), the mathematical properties of these systems can be abstracted and the resulting system subjected to various situations on the computer screen to see what to expect in real life. Simulated wars can be fought (always preferable to the real thing), where probabilities of actions, estimated from real life, determine the outcome. Evolution of populations under various conditions on food supply, climate, etc., can be studied using computer simulation. In

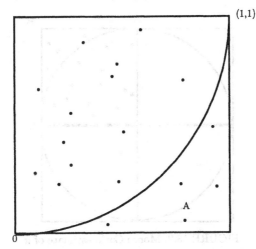

FIGURE 13.1. Monte Carlo estimate of the area under a curve

short, any process can in principle be studied by simulation if you get an appropriate mathematical model of the process and let it evolve in accordance with the built-in probabilities (models of traffic flow, for example, might be related to the Poisson process).

Closely related to simulation are the *Monte Carlo* techniques briefly mentioned before. Monte Carlo methods use simulation to perform calculations. They are frequently based on the Law of Large Numbers, so that the probability of an event is estimated by finding the relative frequency of the number of times the event occurs in a large number of repeated trials.

For an example, recall that the graph of the equation $y = x^2$ in the x-y plane is the type of curve called a *parabola* (see Fig. 13.1). The curve goes through the points (0,0) and (1,1). Consider the unit square of the plane, the square with vertices (0,0), (0,1), (1,0), (1,1). The graph of the parabola $y = x^2$ divides the unit square into two parts, the part above and the part below the parabola. The problem is to calculate the area of the region of the unit square below the parabola. This problem is easily done by elementary calculus, and the solution is beautifully simple: 1/3. Let's see how to estimate this answer using a Monte Carlo method. Choose two random numbers independently, X and Y, from the uniform distribution on the unit interval. It follows from this that the pair (X, Y) is a point of the unit square chosen from the uniform distribution on the unit square, that is, the probability that (X, Y) is in any region of the unit square is just the area of that region. Now let us keep generating these random pairs independently and uniformly, to get the sequence $(X_1, Y_1), (X_2, Y_2), \cdots (X_n, Y_n)$, where each pair represents a point in the unit square, and the points have been chosen uniformly and independently in the square. Define the indicator random variables $Z_i = 1$ or 0 depending on whether (X_i, Y_i) is below the parabola or not. The variables Z_i are independent and identically dis-

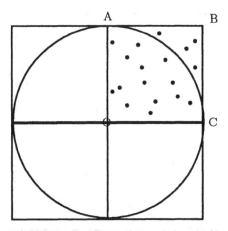

FIGURE 13.2. Monte Carlo estimate of π

tributed, and the ratio

$$\frac{Z_1 + Z_2 + \cdots + Z_n}{n} \tag{13.1}$$

is the relative frequency of the hits below the parabola in the n trials. The Law of Large Numbers assures us that this ratio converges to

$EZ_1 = P((X_1, Y_1))$ lies below the parabola)
$\quad$ = area of the region of the unit square below the parabola.

What this means is that the ratio of formula 13.1 for large n should be a good approximation to the area of the region below the parabola. How large n must be before we can expect the approximation to be within a desired degree of accuracy is the same problem mentioned briefly in Section 8.5, that of studying the rate at which the Law of Large Numbers gets close to its limit.

From a pictorial point of view, we are saying that if you bombard the unit square with lots of independently generated points landing uniformly on its surface, the relative frequency (or proportion) of points landing below the parabola approximates the ratio of the area of the region below the parabola to the area of the entire region. Since the area of the entire region (the unit square) is 1 in this case, the relative frequency approximates the area of the region below the parabola.

As a second example of the Monte Carlo method, let's get an estimate of the number π, the ratio of the circumference of a circle to its diameter. Consider the circle with radius 1 (the units can be anything) and center at the origin (see Fig. 13.2). The area of this circle is π. Notice that the upper right quadrant of the circle lies in the unit square. Let us focus on the unit square and the quadrant of the circle within it. The idea of the procedure is almost the same as in the preceding example: bombard the unit square with points uniformly and independently generated. The proportion of these points falling within the quadrant is an estimate of the probability that a random point falls within the quadrant. This probability is the area of the quadrant, which is known to be $\pi/4$. Therefore, an estimate of π is four times this proportion. You can also estimate π using the Buffon needle approach as we saw in Chapter 12. This was straightforward if you just toss needles, but in using the computer you have to be careful, since in the simulation procedure you are required to choose an angle θ uniformly between 0 and π radians, and doing this requires already knowing π (see Chapter 14 for details). This is bad because you should not need to know the value of π in order to estimate it. You can get around this by using degrees rather than radians, but the Buffon approach to estimating π with the computer is less elementary than the first way.

The Monte Carlo method is valuable in providing estimates for problems where direct calculations may be difficult or impossible to perform. If we wish to calculate the volume formed in space by a number of intersecting surfaces, say, it may be far simpler to use a Monte Carlo procedure similar to the one just described than to attempt other methods. As we mentioned before, there seems to be something strange in using methods depending on chance to estimate explicit quantities having nothing to do with chance. Yet this is perfectly legitimate as long as we realize the limitations of the method. Monte Carlo only gives us estimates, which will in general be different every time we perform a run of the procedure. But if the number of trials is always taken large enough, the results of different runs of the same procedure should be close. A final estimate could be based on an average of a number of runs, each with a large number of trials.

A final important consideration is that Monte Carlo or simulation procedures are only as good as the random number generator you use with them. If the random number generator is poor, don't bet on the results.

13.6   The 1970 draft lottery

We end this chapter by briefly discussing the use of lotteries in the selection of men for military service, in particular, the 1970 draft lottery in the U.S. You can find further details in the article [9].

Lotteries have been used for selection purposes throughout history, where the randomness of the procedure is supposed to ensure fairness. In the U.S., lotteries took place during both world wars for the draft. This involved the random selection of capsules, each containing a number, from a large bowl in public view. The numbers in the capsules corresponded to numbers assigned to men eligible for the draft. The results of the lottery of 1940 were studied by statisticians who discovered serious departures from what one would expect from a truly random set-up. Apparently, the capsules had not been mixed well enough before the drawing. It turns out to be very difficult to mix a large number of physical objects sufficiently well so that the model of randomness can be applied.

In 1970, the lottery was based on birthdays, with each day of the year corresponding to a randomly chosen number from 1 to 366 (this included leap year). Those men with birthdays corresponding to low numbers would be called first. A description of the preparation for the drawing is very interesting in light of the results. First the 31 January dates were each written on slips of paper and inserted into capsules which were then pushed to one side of a large, square wooden box. Then the 29 February dates were placed in capsules and put into the empty part of the box and the January and February capsules were mixed. These were then pushed to one side of the box, and the March capsules were poured into the empty side and then mixed with the January and February capsules, and so on. This procedure clearly does not treat the capsules in a uniform way since the January capsules were mixed with the other capsules 11 times, but the December capsules were mixed with the other capsules only one time. At the drawing, the first capsule drawn was assigned the number 1, the second the number 2, etc., until all the capsules had been chosen and assigned a lottery number in this way. It was noted at the time that the people picking the capsules were seen to choose from the top of the bowl most of the time.

The results of the 1970 lottery were subject to careful statistical scrutiny because of the apparent biases of the 1940 lottery and perhaps partly due to the increasing development and recognition of the field of statistics. The verdict was not good. The 1970 lottery failed a number of statistical tests for data supposedly obtained by random selection. In a plot of the average selection number versus month, a linear trend was found indicating that as the year progressed from January to December, average selection numbers decreased. For a further test the lottery numbers from 1 to 366 were divided into the three groups: 1-122, 123-244, 245-366. It was found that the months were not homogeneously distributed over these groups. In particular, it was observed that the first four months of the year appeared less frequently than the other eight months in the first of these groups, confirming results of the other test. Several more tests added further evidence for the argument that the drawing could not be considered truly random. In addition, the apparent skewness of the results seemed consistent with the way the capsules were prepared for the drawing. Since the capsules corresponding to months later in the year were mixed in with the others less and less, one might expect low numbers to predominate in later months if the top of the box tended to contain capsules from the later months. In fact, December contained 17 numbers from the first group 1-122, whereas January only had 9 numbers from this group.

Because of these problems the lottery of 1971 was more carefully designed (statisticians were actually consulted to help out, which was apparently not the case for previous lotteries). There was a physical mixing of the birthday capsules and then a drawing took place, all in public. But then lottery numbers were not assigned to be simply the trial number at which a birthdate capsule was chosen, as in the previous lottery. Now another drawing was made from another box containing the numbers 1 to 365 which had also been publicly mixed. As each birthdate capsule was chosen from one box, a number was chosen from the other box, and this set the lottery number for the birthdate capsule drawn. But here is the most important innovation from previous lotteries: For both drawings, statistically legitimate random permutations provided by the National Bureau of Standards were used. The random permutations governed *the order* in which the capsules containing the days of the year and the lottery numbers were put into the boxes *before* the physical mixing occurred. In effect, then, the random permutations started mixing everything up before the physical mixing took place. The physical mixing added more to this randomizing process and also served as a public display of what people think of as random.

13.7 Exercises for Chapter 13

1. Outpatients arriving for medication at a hospital are going to be part of a clinical trial testing a new drug against an old one. Upon arrival at the hospital, each patient is placed in either group 1 receiving the new medication or group 2 receiving the old one. This is to continue until a total of 40 patients have been assigned. Describe a way to assign the patients to a group. How would you modify the procedure if each group is to have 20 patients?

2. Describe a procedure using a random number table for choosing digits 0 or 1, each with probability 1/2. Give a similar procedure for choosing digits 0, 1, or 2, each with probability 1/3.

3. Describe the difference between a pseudo-random and a random number.

4. About how often would you expect to see a run of three identical digits in a random number table?

5. A large number of points bombard a square of side 3 units in independent trials, each point uniformly distributed on the surface of the square. Thirty-five percent of the points fall in a certain region R of the square, and the other sixty-five percent fall outside R. Give a Monte Carlo procedure and estimate for the area of the region R.

14
Computers and Probability

You can hear them sigh and wish to die,
You can see them wink the other eye,
At the man who broke the bank at Monte Carlo.
The Man Who Broke The Bank at Monte Carlo
popular song, words and music by Fred Gilbert

14.1   A little bit about computers

As we have seen in the previous chapter, a modern computer with a good
random number generator allows you to do simulations and Monte Carlo
calculations, and in so doing enables you to see many of the fundamental
laws of probability in action. Lots of the results in probability, such as the
Law of Large Numbers, require a large number of trials before something
nice happens. With the computer, we can simulate a large number of trials
in a very short time, so we should be able to test out our results and even
perhaps discover new ones by playing around with probability and the
computer. In this chapter, we want to do a little of this, so some familiarity
with a personal computer and a programming language will be assumed.
We are going to write some elementary algorithms for simulation and Monte
Carlo procedures illustrating examples discussed in preceding chapters. An
algorithm is just a step-by-step description of the solution of a problem in a
finite number of steps. We are going to write the algorithm in pseudocode,
that is, ordinary English. To get these algorithms to produce output on

a computer, you will have to translate the pseudocode into a computer language you are familiar with. I used a language called QBASIC, a version of the well-known programming language BASIC. The execution time of the programs derived from the algorithms given below will depend on a variety of factors, for example, the number of trials requested and the power of your computer.

All of the algorithms below depend on the computer's ability to generate pseudo-random numbers (recall the discussion about generating pseudo-random numbers in Chapter 13). The instruction "set random seed" used in the algorithms means that the random seed determining the sequence of pseudo-random numbers to be used is chosen either by you, by computer default, or preferably by the computer in a quasi-random way. There is an instruction that does this last option, for instance, by letting the time of the day at the instant the instruction is processed set the seed.

The sequence of random numbers produced is usually uniform on the unit interval in the sense that a finite decimal is generated. In QBASIC, such a decimal is produced by the command "Print RND." At execution, this gives a seven-digit decimal (e.g., .7055475), where each of the digits is random.

Many of the algorithms given below need random digits from some finite set, but these can always be obtained from the random decimals given by the computer by using simple transformations. In QBASIC, another way to get a random digit 0 through 9 is to use the expression "INT(10*RND)." This takes the random decimal RND on the unit interval, stretches it to a random decimal on the interval 0 to 9.99+, then selects the integer part of this decimal. Each digit appears with probability .1. A random digit 0 or 1, each occurring with probability .5, can be obtained in QBASIC by setting, for example, $X = 0$ if RND $\leq$ .5 and $X = 1$ otherwise. To translate the following algorithms into proper programming code, you will have to be familiar with the instructions for choosing random numbers of the type desired in the programming language you are using. Please note that instructions below telling you to choose random digits, but not specifying any way to do this, intend the user to choose each of the possible digits that can arise with the same probability. For example, if we are told to choose one of the digits between 1 and 3, give each of them the probability 1/3.

It is also important to mention that in any Monte Carlo procedure you will want to know how big the number of repetitions should be to get a decent approximation to the quantity sought. As we have said before (see, e.g., Section 8.5) there are methods to answer this, but we won't worry about this problem. We will simply take several different values for the number of repetitions and see what happens.

14.2   Frequency of zeros in a random sequence

We're ready to write the first algorithm. The user chooses a number of trials, N. For the first trial, a random digit between 0 and 9 is generated. If the digit is 0, a counter is increased by 1. This procedure is repeated until N trials are obtained. The proportion of zeros is obtained by dividing the value of the counter by N. For large N, the Law of Large Numbers says this relative frequency should be close to .1. By changing the value of X in step 5, the algorithm can be modified to work for any of the other digits.

```
1. Set random seed

2. Enter number of trials N

3. Initialize counter to 0

4. Choose a random digit X between 0 and 9

5. If X=0 then increase the counter by 1

6. Repeat steps 4 and 5 until N digits have been processed

7. Print the proportion of zeros obtained; it is the value
   of the counter divided by N
```

I ran this program for five values of N: 50, 100, 1,000, 5,000, and 10,000, obtaining relative frequencies of .04, .08, .099, .1062, and .1003, respectively. Notice how the approximation to the limiting value of .1 is poor for relatively small values of N and improves for large values. A slight change in the algorithm prints out each number generated, giving us a random number table.

14.3   Simulation of tossing a coin

This algorithm simulates tossing a fair coin. It is very similar to the preceding recipe except that a random digit 0 or 1 is chosen. Besides giving the relative frequency of heads (identified with the digit 0) at each trial, "H" or "T" is printed depending on whether heads or tails was obtained on that trial. From the Law of Large Numbers, the proportion of heads should be close to .5 for large N.

```
1. Set random seed

2. Enter number of trials N

3. Initialize counter to 0
```

4. Choose a random digit X equal to 0 or 1

5. If X=0 then print ''H'' and increase the counter by 1

6. If X=1 then print ''T''

7. Repeat steps 4, 5, and 6 until N digits have been processed

8. Print the proportion of ''heads''; it is the value of the counter divided by N

For 1,000 tosses I got a relative frequency of .517; for 5,000 tosses, .5144; and for 10,000 tosses, .5014 (you might want to suppress the printing of the actual tosses for a large number of trials).

14.4 Simulation of rolling a pair of dice

To simulate rolling a pair of dice, we generate a pair of random digits between 1 and 6. Here we set two counters to store the total number of times 7 and 11 appear; these are used to calculate the relative frequencies of 7 and 11.

1. Set random seed

2. Enter number of trials N

3. Initialize counter7 and counter11 to 0

4. Choose two random digits X and Y, each between 1 and 6

5. If X+Y=7 then increase counter7 by 1

6. If X+Y=11 then increase counter11 by 1

7. Repeat steps 4, 5, and 6 until N pairs of digits have been processed

8. Print the proportion of 7's and 11's; they are obtained by dividing counter7 and counter11 by N

From the Law of Large Numbers, the relative frequency of 7 should be close to .1666··· and that of 11 should be close to .0555····. For three runs using 1,000, 5,000, and 10,000 rolls, I got relative frequencies of .149, .164, and .1621 for 7; and .065, .053, and .0547 for 11. Notice that for 10,000 rolls the proportion for 7 is not as close to the theoretical value of .1666··· as the proportion for 5,000 rolls. Isn't this odd since the Law of Large Numbers tells us that with an increasing number of trials the approximation to the theoretical value should be getting better? No, it isn't odd, because the theory only tells us that most sample runs will average out in the right way in the long run; nothing is said about comparing two different sample runs or, even if we know a sample run averages out right, how many trials it takes to get close to its limit.

14.5   Simulation of the Buffon needle tosses

Let us take a needle of unit length $L = 1$ and with distance between lines $D = 2$. According to the analysis in the preceding chapter and the discussion in Chapter 11, the distance X is chosen uniformly on the unit interval and θ is chosen uniformly on the interval from 0 to π. The needle hits a line if the inequality of step 6 is valid. The relative frequency of number of times the needle hits a line should be close to $1/\pi \approx .3183$.

1. Set random seed

2. Enter number of trials N

3. Initialize counter to 0

4. Choose a random number X uniformly on the unit interval

5. Choose a random angle θ uniformly on the interval 0 to π

6. If $X < \frac{1}{2}\sin\theta$ then increase counter by 1

7. Repeat steps 4, 5, and 6 until N pairs (X,θ) have been processed

8. Print the proportion of times the needle hits the line; this is the value of counter divided by N

I ran this for 500, 1,000, and 10,000 tosses, and I got .302, .334, and .3176, respectively, for the estimates.

14.6   Monte Carlo estimate of π using bombardment of a circle

This was explained in Chapter 13. Random numbers chosen uniformly from the unit interval are chosen in pairs to get points in the unit square. If the inequality of step 5 is valid, the point (X, Y) falls into the upper quadrant of the circle with radius 1 and center at the origin. The relative frequency of points in the circle is an estimate of $\pi/4$; this is used to print out an estimate of π in step 7.

1. Set random seed

2. Enter number of trials N

3. Initialize counter to 0

4. Choose two random numbers X and Y, each uniform on the unit interval

5. If $X^2+Y^2 < 1$ then increase the counter by 1

6. Repeat the steps in 4 and 5 until N pairs of random numbers have been processed

7. Print the estimate of π; this is four times the counter divided by N

This was run for 1,000, 5,000, and 10,000 trials, yielding estimates 3.16, 3.1648, and 3.12, respectively.

14.7   Monte Carlo estimate for the broken stick problem

In Section 11.6, we found the probability that a stick broken at random at two points produces three pieces forming a triangle. The following algorithm gives a Monte Carlo estimate of this probability. For each of the N trials, the stick is broken at two random points, u and v. In the solution of the problem, you recall, there were three inequalities required to hold if a triangle is possible. These form the condition of step 6, and the counter is incremented each time all three inequalities are true.

1. Set random seed

2. Enter number of trials N

3. Initialize counter to 0

4. Choose random numbers u and v, each uniform on the unit interval

5. Set the larger of u and v equal to Y, the smaller equal to X

6. If X < .5 and Y > .5 and Y-X < .5 then increase the counter by 1

7. Repeat steps 4, 5, and 6 until N random pairs (u,v) have been processed

```
8. Print the proportion of times the triangle can be formed;
   this is the value of the counter divided by N
```

There were four runs using 500, 1,000, 10,000, and 20,000 trials. The corresponding estimates of the true probability .25 were .224, .259, .2475, and .2473, respectively.

14.8   Monte Carlo estimate of a binomial probability

Toss a fair coin 10 times. The probability of obtaining exactly three heads is calculated from the binomial distribution (Section 7.2) to be $C_{10,3}\, 2^{-10} \approx$ .1172. We estimate this probability by tossing the coin 10 times in N repetitions and counting the relative frequency of times exactly three heads are obtained. Random digits 0 and 1 are generated, where 1 represents getting a head. If the sum of the 10 tosses is 3 for a repetition, then the counter is incremented by 1.

```
 1. Set random seed

 2. Enter number of repetitions N

 3. Initialize counter to 0

 4. Initialize sum to 0

 5. Choose a random digit X equal to 0 or 1

 6. Add the value of X to sum and store the result in sum

 7. Repeat steps 5 and 6 until ten random digits have been
    processed

 8. If sum=3 then increase the counter by 1

 9. Repeat steps 4 through 8 until N repetitions have been
    processed

10. Print the estimate of the probability of exactly three
    heads in ten tosses of a fair coin; this is the value of
    the counter divided by N
```

The estimates were .123, .1142, and .1199 from 1,000, 10,000, and 15,000 trials, respectively.

Other binomial probability estimates can be obtained with slight variations of this algorithm. If, for example, the success probability is .1 for a

Bernoulli trial set-up, the above algorithm can be easily modified to estimate the probability of obtaining exactly three successes in ten trials. Let
success be equivalent to selecting the digit 1 with probability .1. Then step
5 above should be replaced by:

```
5. Choose X equal to 0 or 1 with probabilities .9 and .1
```

This can be done by selecting a random digit U equal to 0 through 9, and
then letting $X = 1$ when $U = 1$, say, and $X = 0$ whenever U is equal to
any digit other than 1. The binomial distribution gives an exact value of
the probability to be $C_{10,3}\,(.1)^3\,(.9)^7 \approx .0574\cdots$. The estimates from runs
of 1,000, 10,000, and 15,000 trials were .043, .0654, and .056, respectively.

14.9 Monte Carlo estimate of the probability of winning at craps

In this program, we simulate N craps games, where N is selected by the
user. A tally is kept of the number of games won. The probability of winning
at craps is then estimated by calculating the relative frequency of games
won. At the start of a game, $X + Y$ gives the initial roll of the dice. If the
game is won, the counter is increased and a new game starts, if lost a new
game starts, and if a point is made we repeatedly roll the dice to get score
$U + V$ until we either get 7 (lose) or the point (win). If the point is won,
the counter is increased by 1. After the point is lost or won, we start a new
game.

```
1. Set random seed

2. Enter number N of games to be played

3. Initialize counter to 0

4. Choose random digits X and Y, each between 1 and 6

5. If X+Y=7 or 11, then increase counter by 1 and go to
   step 12

6. If X+Y=2, 3, or 12 then go to step 12

7. If X+Y=4, 10, 5, 9, 6, or 8, then set point=X+Y

8. Choose random digits U and V, each between 1 and 6

9. If U+V=7 then go to step 12

10. If U+V=point then increase counter by 1 and go to
    step 12
```

11. If U+V ≠ both 7 and point then go back to step 8

12. Go back to step 4 (new game) and repeat until N games have been played

13. Print the proportion of games won; this is the value of the counter divided by N

For four runs of 1,000, 5,000, 10,000, and 20,000 games, I got estimates of .476, .4806, .4916, and .4867, respectively. As we recall from Chapter 6, the exact probability of winning at craps is about .4929.

14.10   Monte Carlo estimate of the gambler's ruin probability

Now we'll make a Monte Carlo estimate of the gambler's ruin probability derived in Chapter 10. Let N be the number of repetitions of the game. The gambler's initial capital, i, and total capital a are entered, as well as the probability q of the gambler losing \$1 (moving one unit to the left). We can represent this by selecting one of the numbers -1 and 1 with probabilities q and p, respectively. If, say, $q = 6/11$, we can let X be a random digit between 1 through 11 and then set $win = -1$ when $X = 1$ through 6 and $win = 1$ otherwise. A counter is initialized to 0 at the outset. Each repetition starts by initializing a sum counter to i, and adding on the value of win at each play of the game until $sum = 0$ (gambler's ruin) or $sum = a$ (adversary's ruin). If the gambler is ruined, then the counter is increased by 1. We repeat for N repetitions.

1. Set random seed

2. Enter number of repetitions N

3. Initialize counter to 0

4. Store the gambler's initial capital i and the total capital a

5. Initialize sum to i

6. Choose a digit win equal to −1 or 1 with probabilities q and p, respectively

7. Add win to sum and store this value in the variable sum

8. If sum=0 then increase counter by 1 and go to step 11

9. If sum=a then go to step 11

10. If sum ≠ both 0 and a, repeat from step 6

11. Go back to step 5 (new game) and repeat until N
    repetitions have been processed

12. Print the estimate of the gambler's ruin probability;
    this is the value of the counter divided by N

A run of 5,000 repetitions for $i = 2$, $a = 5$, and $q = .5$ gave an estimate of .598 for the true value of .6, and another run of 5,000 for $i = 3$, $a = 5$, and $q = .8$ yielded the estimate .9424 for the true value of .9384.

14.11   Constructing approximately normal random variables

This exercise shows how we can construct random variables having an approximately standard normal distribution by using the Central Limit Theorem discussed in Section 12.3. We are going to add up 900 independent indicator variables X, each having the values 1 and 0 with probability .5 (so you can think of the ith variable as the indicator of head on the ith toss of a fair coin). It turns out that the approximation to the normal distribution when you have binomial variables may not be very good if you don't have a lot of terms to add up; that's why we are taking so many variables in this exercise. Now we standardize the sum by subtracting the expectation of the sum and dividing by its standard deviation to get a variable Y. According to the Central Limit Theorem, Y should have an approximate standard normal distribution since the number of summands is large. To check this very roughly, we make a Monte Carlo estimate of the probability of Y falling in the interval from -1 to 1. If indeed the distribution of Y is close to the standard normal, this probability should be close to the probability .68 that a standard normal variable Z lies in the same interval. The estimate is made by calculating Y N times and increasing a counter each time that Y lies in the interval.

1. Set random seed

2. Enter number of repetitions N

3. Initialize counter to 0

4. Initialize sum to 0

5. Choose a random digit X equal to 0 or 1

6. Add X to sum and store this value in sum

7. Repeat steps 5 and 6 until 900 digits X have been
   processed

8. Let Y=(sum-450)/15

9. If Y > −1 and Y < 1 then increase the counter by 1

10. Go back to step 4 and repeat until N repetitions have
    been processed

11. Print the estimate of the probability of Y falling in
    the interval from −1 to 1; this is the value of the
    counter divided by N

For three runs of 100, 300, and 1,000 trials, the estimates were .62, .6633, and .678, respectively.

You can try doing this exercise where you substitute 50 or 100 for 900 in step 7. How good is the estimate now?

In this chapter, I have tried to give you a taste of simulation and Monte Carlo procedures. I've also attempted to get across how much fun it is to combine the computer with probability ideas. I encourage you to think of other investigations into simulation and Monte Carlo estimation to develop on your own. A few ideas along these lines are given in the exercises below.

14.12   Exercises for Chapter 14

1. (a) Write a simulation for the game of chuck-a-luck (Section 7.3). (b) Refine the simulation in (a) to make a Monte Carlo estimate of the probability of winning $2 at chuck-a-luck.

2. Write a simulation for the car-goat game of Chapter 1, and then use it to estimate the probability of winning if you switch doors.

3. Modify the algorithm in section 14.11 to see whether each of the events $Y > .5$ and $-.3 < Y < .3$ are reasonably close to the values

$$P(Z > .5) \approx .3085 \text{ and } P(-.3 < Z < .3) \approx .2358$$

for Z a standard normal variable.

4. Write a simulation of the game of roulette, and then estimate the probability of winning by playing black.

5. Make a Monte Carlo estimate of the area between the curves $y = x^2$ and $y = x^3$ within the unit square (this is the square with lower left-hand corner the origin and with length of side 1 unit).

6. Rosenkrantz and Guildenstern play the following game: There are 100 coins numbered from 1 to 100. Coin number k has probability $1/k$ of falling heads. A coin is chosen at random and tossed. If the coin has an even number, then R. pays G. $1 if heads comes up, otherwise G. pays R. $1. If the coin has an odd number, then G. pays R. $1 if heads comes up, otherwise R. pays G. $1. Write a simulation of this game and use it to estimate the probability of G. winning $1 on a single play of the game.

15

Statistics: Applying Probability to Make Decisions

> Thou shalt not sit;
> With statisticians nor commit;
> A social science.
> **W.H. Auden,** *Under Which Lyre*

15.1 What statistics does

Whereas probability can be loosely described as the mathematical theory of *measuring* uncertainty, the discipline of statistics can be roughly characterized as the mathematical theory of *making decisions* in the face of uncertainty. In previous chapters, we have already seen some instances of statistical problems (for example, deciding whether a finite sequence of numbers is random). The main theoretical tool of the statistician is probability: in order to make rational decisions, it is necessary to measure the uncertainty for different possible outcomes. The statistician is primarily interested in learning about an entire population of things by studying a small sample of those things and then making inferences about the entire population from the evidence gathered from the sample. What the statistician does is called *inductive inference*, reasoning from the particular to the general case. As you can see, such an activity is fraught with danger.

When the statistician makes a decision, she knows the decision may be a wrong one; being in the uncertainty business means that the possibility of error must play an important role in her procedures and methods. The job of the statistician is to devise an optimal way to make these decisions from a rational point of view, using the scientific method. For this reason, the statistician is focused on *observation*. We start off with simple examples showing how careful you have to be when trying to interpret raw data, and how the inappropriate presentation of numbers can be used, consciously or not, to deceive or arrive at dubious conclusions.

15.2   Lying with statistics?

A famous quip says there are lies, damn lies, and then there are statistics. The implication is that you can support any assertion if you present data selectively. There are valid and invalid ways to consider data, and invalid ways are often used, consciously or not, to try and support dubious claims. To take a few crude examples, suppose it is claimed that 8 out of 10 people prefer product A to product B. This may sound impressive until you discover that 100 people were polled with 8 of them preferring product A, 92 of them product B. The hucksters merely ignored 90 percent of the data that was not to their liking. In another instance, it is claimed that five times as many men as women regularly watch a certain television program. A look at the data shows that 100 men and 10 women who watch television at that hour were polled, with 20 men and 4 women watching the program. The interpreters of the data used the raw scores, 20 and 4, as the basis of their statement. These raw scores are, however, meaningless; what is needed is an estimate of the *proportion* of men and women who watch the program. The proportions of men and women favoring the program are 20 and 40 percent, respectively. From this perspective, it could be argued that twice as many women as men prefer the program. But this conclusion is also dubious since so few women were polled.

More subtle dangers lurking in data often show up in so-called *confounding* factors. A well-known study of sex bias in admission to a graduate school showed that 44 percent of male applicants were admitted compared to only 35 percent of female applicants. This seemed to show bias against women. However, admission to the school was made on a department-by-department basis, and looking at the departments separately there did not appear to be any bias. It turned out that many women had applied to the harder departments to get into and many men had applied to the easier ones. The different preferences of department by men and women had suggested a spurious conclusion about sex bias in the overall statistical summary. This association vanished once researchers controlled for the variable of differing departments.

Descriptive statistics is the art of presenting the data in a clear, informative way, in charts or tables, graphs, etc., with an eye toward making the important points stand out. As we have seen, poor or invalid data presentation can be misleading and is often used when there is a strong desire to persuade, as in advertising or political agendas. Let's now go to problems where the crude presentation of the data is not the issue; it's the more subtle mathematical analysis used in formulating statistical inferences. We'll start making these ideas concrete with a basic problem in testing hypotheses.

15.3   Deciding between two probabilities

One of the main questions of statistics involves judging whether a proposed mathematical model for phenomena is so out of line with direct observations of the phenomena that it should be rejected in favor of a competing model. Suppose, for example, your friend is a famous gambler. He has in his possession an unfair coin which has been constructed so that the probability of falling heads at each independent toss is $3/4$ rather than $1/2$. He presents you with a coin and challenges you to tell whether the coin is fair or is the unfair one. You know exactly one of those two situations is true, and must decide in favor of one of them on the basis of knowing how many heads occurred in 100 (independent) tosses of the coin. Let us describe the two possible situations as two *hypotheses*, H_0 for $p = 1/2$, and H_1 for $p = 3/4$. Now suppose the coin is tossed and you observe 80 heads in the 100 tosses. How should you decide?

Even someone with no knowledge about calculating with probabilities could give a rational argument for deciding in favor of H_1 in this case. The argument could go something like this: under H_0 we expect roughly half the tosses, around 50, to be heads. But under H_1 we expect many more heads, around 3 out of 4 tosses on average, giving 75 heads. The observed value 80 is so far from the expected value of 50 under H_0 that we reject this hypothesis as too unlikely under the circumstances. What we observed was more compatible with H_1 since 80 is closer to 75 than to 50.

Similar gut-level thinking which seems intuitively correct is also possible if you had observed 42 heads, say. In this case, it appears reasonable to decide in favor of H_0 because 42 is so much more in the neighborhood of the expected value 50 under H_0 than in the neighborhood of 75 under the competing hypothesis H_1. But a full solution of the problem requires us to make a decision *for all possible* observations of numbers of heads in 100 tosses. A decision algorithm telling us what to do in all possible cases is called a *test of the hypothesis H_0 against the alternative H_1*. What should we do, for instance, if we had observed 63 heads? Here intuition fails and we need some hard-nosed general principles to put to work.

In order to state such principles, we must fully understand what is expected of us according to the formulation of the problem. Suppose we observe 0 heads occurring in 100 tosses. This observation might well make us seriously doubt the truth of *both* proposed hypotheses H_0 and H_1, since 0 heads would be exceedingly unlikely under either of these models. But if the ground rules of the problem tell us, as they do here, that we must choose precisely one of the competing alternatives, the more rational choice for 0 observed heads is clearly H_0, for 0 is closer to an expected value of 50 than to an expected value of 75. Therefore, if we must choose one of the competing alternatives, a reasonable approach to the general solution can be formulated as follows: there is a critical value $c, 0 \leq c \leq 100$, such that if we observe h heads we should accept H_0 whenever $h \leq c$ and reject H_0 whenever $h > c$. (Rejecting H_0 is the same as accepting H_1—by tradition statisticians usually speak in terms of accepting or rejecting H_0, called the *null hypothesis*). What we have done here is develop the general idea that if we have to make a choice between a relatively small probability of heads and a larger one, we should decide in favor of the smaller one when we observe a relatively small number of heads and in favor of the larger one when we observe a relatively large number. The distinction between relatively small and relatively large will be made by determining a cut-off point c.

The natural question to ask at this point is: how do we determine c? The answer is that there is no way to determine c uniquely without introducing further criteria, some kind of statistical principles to guide us. One such principle for determining c is the *method of maximum likelihood*. It asserts that you should choose the alternative that maximizes the probability of the event actually observed. That is to say, if you observed h heads, compute the probability of obtaining h heads under each of the hypotheses H_0 and H_1 and choose the hypothesis giving you the larger probability. Let's see what this gives us in the general case in which we observe h heads in the 100 trials. The probabilities of observing h heads under H_0 and under H_1 are, respectively,

$$C_{100,h}(.5)^h(.5)^{100-h} \text{ and } C_{100,h}(.75)^h(.25)^{100-h}.$$

The principle of maximum likelihood has us reject H_0 whenever the first probability is smaller than the second. This is equivalent to rejecting H_0 when

$$(.5)^h(.5)^{100-h} < (.75)^h(.25)^{100-h}.$$

By taking logarithms, you get the equivalent inequality

$$\frac{100(\log(.5) - \log(.25))}{\log(.75) - \log(.25)} < h.$$

Use a caculator to evaluate the left-hand side; you get approximately 63.1. What we have shown is that whenever the number of observed heads h is less than or equal to $c = 63$ we should accept H_0; whenever h exceeds 63, we should reject H_0.

But now suppose we had wanted to test $H_0 : p = 1/2$ against the alternative $H_1 : p = 1/4$ rather than $p = 3/4$. In this case, intuition suggests that a rational test is given by rejecting H_0 when we observe *too few* heads in 100 tosses, where a cut-off point can once again be determined according to the maximum likelihood method. In the original problem, the rejection region (for H_0) is a "right tail" one; now it will turn out to be a "left tail" one (a "tail" is an interval of values either greater than some constant or less than it). To see this, we must compare

$$C_{100,h}(.5)^h(.5)^{100-h} \text{ and } C_{100,h}(.25)^h(.75)^{100-h}.$$

Rejection of H_0 occurs this time when

$$(.5)^h(.5)^{100-h} < (.25)^h(.75)^{100-h},$$

and this leads to the relation

$$\frac{100(\log(.5) - \log(.75))}{\log(.25) - \log(.75)} > h.$$

Notice that the direction of the inequality has been reversed; that's because the denominator (the bottom expression in the fraction on the left side) is negative, and we had to divide by this to get h alone on the right. This relation is solved to get $h < 36.9$, approximately, so that H_0 should be rejected whenever there are 36 or fewer heads and accepted otherwise.

15.4 More complicated decisions

The essential statistical idea of the method of maximum likelihood is this: if the observation we saw is more likely to arise under hypothesis A than under an alternative hypothesis B, then decide in favor of hypothesis A. We may, of course, be making an error: hypothesis B may in fact be true. The decision is based on the empirical fact that nature shows us the more probable more frequently than the less probable, so we judge what we observed to be typical. We can carry the rationale underlying the method of maximum likelihood further. The method itself depends only on knowing which of the alternative hypotheses gives a larger probability for what we observed. We want to refine this now by using the actual values of these probabilities to measure our confidence in the decision. For example, there is more confidence in the decision to accept hypothesis A the larger the probability of the observation is under A and the smaller it is under B. In this section, we want to consider applying these ideas to some more complicated problems of testing hypotheses.

Your friend, the gambler, offers you a proposition. He will provide a coin which will be tossed in independent trials. Every time the coin comes down heads you pay him a dollar; every time it comes down tails he pays you a dollar. Before you agree to play, you ask to test the coin for fairness. Your friend reluctantly agrees; he can't understand why you don't trust him—he claims that his history of always winning such games with others is just chance. You ask him whether he will use the same coin but let *you* win if the coin falls heads and lose if it falls tails. He refuses, maintaining that the coin is fair, but because he has always been lucky playing for heads, he won't allow you to play for it. But he will allow you to toss the coin 100 times to check it out. Suppose you do this and get 60 heads. Should you play?

Let p be the probability of the coin falling heads on each toss. To decide whether the coin is fair (and therefore whether to play) you can set up the null hypothesis $H_0 : p = 1/2$ versus the alternative $H_1 : p > 1/2$. If H_0 is true, the coin is fair and you will play, but if H_1 is true the coin is biased in favor of heads and your friend's fortunes. To test H_0 against H_1, the same intuitive principle as in Section 15.3 qualitatively describes what should be done: reject H_0 if you observe more than a certain critical number c of heads, because the more heads obtained the more likely it is that p is larger than $1/2$. But you cannot use the method of maximum likelihood as in Section 15.1 to determine c because the alternative H_1 does not uniquely specify a distribution to compete with the one specified by H_0.

Instead, try the following attack. Let $h =$ number of heads observed in the 100 trials, and calculate $P(h \geq 60)$ *under the null hypothesis*. This is called the *p-value* of the null hypothesis (the p in p-value has nothing to do with the probability p of the coin falling head). The p-value is the probability of observing a value at least as large as the value actually observed if the null hypothesis is true. One way to think about the p-value is to regard it as a measure of how surprised we are at our observation assuming the truth of the null hypothesis. If the p-value is very small, we should be very surprised at getting such an observation under the null hypothesis and take this as evidence supporting the alternative H_1 since the observed value would be unlikely under H_0. On the other hand, if the p-value is large, then we are not very surprised at our observation; since there is large probability of observing values even larger than the one we saw, our observation does not provide evidence for the rejection of the null hypothesis. How small should the p-value be before we reject H_0? That is discretionary and depends on how strong you want the evidence to be before you decide to reject. Traditionally, statisticians have required that p-values be smaller than or equal to .05 before they reject H_0. A p-value of .05 means that you will observe a measurement at least as large as the one actually recorded only 5 times out of 100 if H_0 is true. The smaller the p-value, the stronger the evidence supporting rejection of the null hypothesis, so that a p-value of .001 is extremely convincing since, if the null hypothesis were true, we would see readings at least as large as the one actually observed only 1 in 1000 times. In our problem, for example, if you really trust your friend you may want rather strong evidence that the coin is biased in his favor before you reject the assumption of fairness (the null hypothesis), so you may decide to reject only if the p-value is at most .01 or .001. On the other hand, if you are very fearful of losing money in a biased game, you might require very little evidence of bias before deciding to reject the null hypothesis, so you would take the p-value to be .05 (or even greater).

How do we actually find the p-value in this problem? We would need to calculate

$$P(h \geq 60) = P(h = 60) + P(h = 61) + \cdots + P(h = 100)$$

using the binomial distribution probabilities. The computation is cumbersome because the binomial coefficients $C_{100,x}$ must be calculated and added up for $x = 60, 61$, etc. There are tables of the binomial distribution which could help you out here. You can also find estimates of the binomial terms. Another way is to get an approximation by using the Central Limit Theorem. Let X_i be the indicator functions with values 1 or 0 depending on whether the ith toss gives head or tail. If H_0 is assumed true,

$$Z = \frac{X_1 + X_2 + \cdots + X_{100} - 50}{5} \qquad (15.1)$$

would have an approximately standard normal distribution if 100 is large enough for the convergence to the normal to be good (see Chapter 12). Unfortunately, however, for a binomial distribution like this, the approximation using the normal distribution may not be so good with only 100 observations (see also Section 14.11). It would be better to take 1000 observations, say. But let's pretend that the approximation is better than it is for the sake of illustration. The sum of the X's is just h, so $P(h \geq 60)$ is by formula 15.1 equivalent to $P(Z > 2)$. From a table of the normal distribution, we see this probability, the p-value, is between .02 and .03. If this value is used as an approximation to the p-value of the binomial distribution, you have reasonable evidence to reject H_0 and refuse to play. As noted, there is a problem because the sample size 100 is small for the approximation to be reliable—after all, if the binomial probability differs from the normal probability by as little as .03, it could change our decision from rejection to acceptance of H_0. But the idea of using the normal approximation is a good one provided the approximation is good enough, and this will be so provided the sample size is sufficiently large.

A similar approach can be taken if the form of the alternative hypothesis were $H_1 : p < 1/2$ or $H_1 : p \neq 1/2$. In the former case, rejection of the null hypothesis should occur if too few heads are observed (a left-tail rejection region). The p-value should now be calculated as $P(h \leq a)$, where a is the observed number of heads, and, as before, this probability can be approximated (again poorly, with only 100 observations) using the standard normal distribution. In the latter case, rejection of the null hypothesis should occur if either too many or too few heads are observed (a two-tail rejection region). In this case, we should consider $P(h \geq a)$ and $P(h \leq a)$ where a is the observed number of heads, and let the p-value be twice the smaller of these. What you are doing here is using the smaller number to define a rejection region for one tail and then using the same number to define a rejection region for the other tail. The total probability of rejecting H_0 is then the p-value.

To illustrate, suppose you want to decide whether a coin is fair or not fair. Unlike the previous problem involving your gambler friend, in which the alternative to the null hypothesis was that the coin is biased in a particular direction, now you want to reject the null hypothesis if you have evidence of bias in either direction. Accordingly, we set $H_1 : p \neq 1/2$. If 60 heads are observed, the p-value is twice the right-tail p-value obtained before, namely, some value between .04 and .06. In this case, the evidence against the null hypothesis is less convincing than before since the p-value is rather large. This is not surprising on intuitive grounds; if you don't know the direction of the possible bias, there are more kinds of evidence that can turn up supporting H_1. Therefore, any particular piece of evidence should have less weight than in the one-tail situation to keep the total weight of evidence the same.

15.5 How many fish in the lake, and other problems of estimation

We want to estimate the number of fish in a lake. To do this we catch 1000 fish, tag them, and release them back into the lake. After a day or two, we get another catch of 1000 fish and an examination finds that 200 of these are tagged. What's a reasonable estimate of the fish population in the lake?

One approach to this problem harks back to random bombardment of an area by an independent, uniformly distributed set of points. This was discussed in Chapters 13 and 14 with reference to Monte Carlo estimation techniques. The main idea was that the ratio of the number of points in any region relative to the number of points in another region should be roughly the same as the respective ratio of the areas of the two regions. So if we toss N points onto a unit square and i points fall into a region of unknown area x, then x may be estimated by i/N, where the estimate becomes better the larger N is.

Now suppose the unit square represents the lake and the 1000 tagged fish play the role of the bombarding points. Instead of measuring mass by area, let the mass of any region be measured by the number of fish in it. The entire lake has an unknown number of fish, N. A few days after the tagged fish are released, go to a subset S of the unit square (a part of the lake) and count the number of points in it (count the 200 tagged fish), and also the total mass of S (total of 1000 fish in this part of the lake). The conclusion is that the proportion $200/1000$ of tagged fish in S should be approximately equal to $1000/N$, the proportion of tagged fish in the whole lake under the assumption that the tagged fish have uniformly dispersed in the lake. This gives an estimate of 5000 fish in the lake. Of course, for this approach to work, the fish in the lake will have to be modelled reasonably well by the uniform, independent distribution of points in a plane. We must assume that the tagged fish will mix well, spreading uniformly through the water, and the fish population will not change appreciably between the two catches. Our estimate is a single number, called a *point* estimate. A more useful kind of estimate is an *interval* estimate, or *confidence interval*, which we'll talk about later.

Another way to look at the problem involves considering the probability $p(200, N)$ that a catch of 1000 fish in a lake having N fish yields exactly 200 tagged fish from a total of 1000 tagged fish. The total number of ways of choosing 1000 fish from the N fish in the lake is $C_{N,1000}$. Now think of what we actually caught—we got 200 tagged fish from a total of 1000 tagged fish, which can be done in $C_{1000,200}$ ways, and then 800 untagged fish from a total of $N - 1000$ untagged fish, and this can be done in $C_{N-1000,800}$ ways. Using the ideas of uniform probabilities on finite sample spaces, we get

$$p(200, N) = \frac{C_{1000,200} \cdot C_{N-1000,800}}{C_{N,1000}}.$$

(If we wrote these probabilities down for all possible values of caught tagged fish from 0 to 1000, we would get an example of what is called a *hypergeometric* distribution.) At this point, we can invoke the principle of maximum likelihood again. We observe that $p(200, N)$ is a function of N, and that a reasonable estimate of N would be the value of N making $p(200, N)$ as large as possible. It is not very hard to find the value of N maximizing the expression on the right-hand side of the above relation, but we will spare you the details here. The answer comes out the same as before: use 5000 fish as your estimate. The same restrictions on the fish in the lake for this model to apply are the same as for the previous model. That the two methods agree should cause no surprise since we are using much the same reasoning. The assumption of well mixing of the fish population takes the mathematical form of imposing a uniform distribution in both approaches.

The "tagged catch" method has been used to take a census of fish and animal populations. On the other hand, the census of people in the United States has historically been non-statistical in nature—the attempt has been to count each and every person. The trouble is that when a population is so large and complex, counting becomes a formidable task. It has been maintained that, for one reason or another, a substantial number of people are not getting counted. Such errors have serious consequences since the census determines the amount of political power a region possesses. Is it possible that a statistical type of census might not be cheaper, easier, and more accurate than the current method? Roughly, this could be done by randomly selecting a number of regions of different kinds (e.g., urban, suburban, rural) and then counting as accurately as possible the people in these regions. The data obtained from the random samples would then be used to provide appropriate estimates.

An important, general problem in statistical estimation goes like this: we are given a random variable X with some unknown probability distribution. We want to estimate some of the basic parameters of this distribution, like the expectation of X and the variance of X. First let's see how to estimate $EX = \mu$. The standard way to do this is to take a random sample from the distribution of X; that is, we plan to observe n independent variables $X_1, X_2, \cdots, X_n$ all with the same distribution as X. (The question of how to actually *obtain* a random sample can pose daunting problems in its own right, but for the moment we won't discuss this.) We then consider the random variable

$$\overline{X} = \frac{X_1 + X_2 + \cdots + X_n}{n},$$

the sample mean. This random variable is an example of what statisticians call a *statistic*, namely, a function of the observations not depending on any unknown parameters—when we plug in the value of the observations we must get a number. This certainly happens with $\overline{X}$, which is simply the average value of the observations. A statistic is used when you want to estimate a parameter; in this case, $\overline{X}$ will be used to estimate the unknown value of μ. Often a statistic used to estimate a parameter is called an *estimator*.

Here's a simple example related to Section 15.3. You recall that we considered a situation in which a coin was tossed 100 times and the problem was to make a decision between two hypotheses: whether the coin is fair or whether p, the probability of head, equals 3/4. Let us now change the point of view: instead of testing a hypothesis, suppose we only want to estimate p. The observations consist in tossing the coin 100 times and letting X_i be the usual indicator variables for the tosses, with value 1 or 0 depending on whether we see head or tail on toss i. Notice that $\mu = EX_i = p$, and $\overline{X}$ is just the relative frequency of heads in the 100 tosses. If 63 heads come up, then we are saying that .63 gives you a point estimate of p.

The rationale for using $\overline{X}$ as an estimator of μ is rooted in our old friend, the Law of Large Numbers. It should be clear that our present notation $\overline{X}$ is exactly the same thing as S_n/n (statisticians, by tradition, use certain notations as standard). As you recall (from Section 8.5), we observed that $ES_n/n = \mu = EX_1$, where S_n is the sum of n independent, identically distributed random variables. Therefore $E\overline{X} = \mu$, and the Law of Large Numbers assures us that the sample mean will be concentrated around μ as closely as we want provided n, the number of observations, is made large enough. In fact, if the variance of the X_i is σ^2, a finite number, then by formula 8.16, we have that the variance of $\overline{X}$ is σ^2/n, and therefore $\overline{X}$ converges to the unknown μ in probability. (This is just a restatement of the (Weak) Law of Large Numbers.)

We note above that the estimator $\overline{X}$ has the property that $E\overline{X} = \mu$; that is, the expected value of the estimator $\overline{X}$ is μ, the parameter we are trying to estimate. In general, it seems reasonable to require any statistic T that estimates some parameter λ of a distribution to have the property $ET = \lambda$, that is, it seems reasonable to require a "good" estimator to have expected value precisely the quantity we're trying to estimate. Such estimators are called *unbiased*. An unbiased estimator has a distribution whose average value is the parameter being estimated. Now if we have a sequence of unbiased estimators T_n of λ such that the variance of T_n converges to 0, we have the extremely pleasant situation of the estimators converging to λ in probability—this is what is happening with $\overline{X}$ and μ above. When this happens, we say that the sequence of unbiased estimators is *consistent*. A consistent sequence of unbiased estimators not only has the property that each estimator in the sequence has expectation equal to the parameter being estimated, but as you go out further in the sequence the distribution of the estimators cluster more and more closely around the unknown parameter with probability approaching 1. So for a large sample size, the estimate is very likely giving you a good approximation to the parameter.

Now suppose we want an unbiased estimate of the variance σ^2 of X, which we assume is a finite number. If we know the value of the expected value μ, then we can reason as follows: the variance is $E(X - \mu)^2$, and if we set $Y = (X - \mu)^2$ then the problem is reduced to finding an unbiased estimate of EY. But we have just solved this by taking $\overline{Y}$. Therefore, an unbiased estimator of σ^2 is

$$T = \frac{(X_1 - \mu)^2 + (X_2 - \mu)^2 + \cdots + (X_n - \mu)^2}{n}.$$

In addition, T considered for each possible sample size n turns out to give a consistent sequence of estimators of σ^2.

But what happens if we don't know the value of μ and want to estimate σ^2? We cannot use T since it would depend upon the unknown parameter μ and would not be a statistic. Since μ is unknown, let's try to substitute the estimate $\overline{X}$ for μ in T and see what happens. We get

$$T_1 = \frac{(X_1 - \overline{X})^2 + (X_2 - \overline{X})^2 + \cdots + (X_n - \overline{X})^2}{n}.$$

T_1 is certainly an estimator since it only depends upon the observations, but is it unbiased? We can find out by a tedious computation which I omit; essentially you just have to expand the right-hand side of the above relation for T_1 and take expectations to get lots of terms like EX_i^2 and EX_iX_j for $i \neq j$. The first expectation is $\sigma^2 + \mu^2$ and the second (by independence) is μ^2. Then you have to add up lots of stuff. What you get when you are done is $\sigma^2 \cdot (n-1)/n$, so that T_1 is *not* unbiased. It is, however, very easy to get an unbiased estimator for the variance: use $s^2 = T_1 \cdot n/(n-1)$; that is, replace the n in the denominator in the definition of T_1 by $n-1$. The estimator s^2 is called the *sample variance*; as n becomes large, the difference between T_1 and s^2 becomes negligible.

15.6   Polls and confidence intervals

One of the results of a poll of 1154 adults printed in the *New York Times* of February 16, 1993 (see [7]) claimed that 56 percent of those sampled thought it would be necessary to cut back government programs that benefit people like themselves in order to reduce the budget deficit. In a small box near the article, it explains that "in theory, in 19 cases out of 20 the results based on such samples will differ by no more than three percentage points in either direction from what would have been obtained by seeking out all American adults." What we'd like to do now is learn a little about what polls can tell us and what the quoted remark means.

Suppose the sample consists of n individuals, each of whom is asked a question requiring a "yes" or "no" answer. Define the n indicator variables X_i equal to 1 or 0 depending on whether the ith individual gives a "yes" or "no" response. Then $\overline{X}$ is an unbiased estimator of $EX_1 = p$, where p can be interpreted as the probability that a respondent answers "yes" to the question. The value p can be viewed as the proportion of voters who would have answered "yes" had we polled the entire population, so essentially our problem is to get a decent estimate of p from the sample. For the problem above, we already know .56 is an unbiased point estimate. But point estimates have an intrinsic problem: they pinpoint a single value in an infinite set of possibilities. Every time you take a sample, you will almost certainly get different values for the estimator, so the chances that your estimate agrees with the true value of p are null. We want to introduce the idea of an *interval* estimate of p. What this does is replace the attempt to hit p exactly on the head using a point estimate with an interval estimate which you believe, with a certain degree of confidence, contains the true value of p. To be more precise, consider formula 12.1 and divide the top and bottom of the fraction by n. You get

$$W = \frac{\overline{X} - p}{\sqrt{\frac{p(1-p)}{n}}}$$

but haven't changed the value of the fraction, which still has an approximate standard normal distribution for large n. So we have

$$P(-1.96 < W < 1.96) \approx .95 \qquad (15.2)$$

and by substituting into this relation the right-hand side of the previous relation and then doing a little algebraic rearranging, we get

$$P\left(\overline{X} - 1.96\sqrt{\frac{p(1-p)}{n}} < p < \overline{X} + 1.96\sqrt{\frac{p(1-p)}{n}}\right) \approx .95.$$

What we have done is sandwich p between two quantities. Notice that these sandwiching quantities depend upon p, which means these quantities are not statistics. This is not good for us since we want p to be sandwiched by *numbers* after we make our observations. To fix this up, we will replace p and $1 - p$ by the estimators $\overline{X}$ and $1 - \overline{X}$ to get

$$P\left(\overline{X} - 1.96\sqrt{\frac{\overline{X}(1 - \overline{X})}{n}} < p < \overline{X} + 1.96\sqrt{\frac{\overline{X}(1 - \overline{X})}{n}}\right) \approx .95.$$

We are assuming here that the replacement does not significantly alter the probability of the event, and this is true if n is reasonably large. Now p is sandwiched between two statistics, and the *random interval*

$$I_{\overline{X}} = \left(\overline{X} - 1.96\sqrt{\frac{\overline{X}(1 - \overline{X})}{n}}, \ \overline{X} + 1.96\sqrt{\frac{\overline{X}(1 - \overline{X})}{n}}\right) \qquad (15.3)$$

contains p with probability around .95. Another way to say this is: p will be contained in $I_{\overline{X}}$ about 95 out of 100 times we calculate the sample mean from observations based on a random sample.

Now suppose we take n observations and substitute the values of n and the sample mean into formula 15.3 to get numbers a, b and therefore an interval $I = (a, b)$. This interval is called an approximately *95 percent confidence interval* for p. Because of common misconceptions, it's important to emphasize what precisely we are claiming about the interval I. In the classical theory, which we are discussing here, the unknown parameter p is a *number*, not a random variable, so p is either in I or outside it, and it is meaningless to speak of the probability of p lying in I (the Bayesians, on the other hand, consider p a random variable—see Section 15.7). The expression *95 percent confidence interval* refers to the *procedure* through which I was produced. This procedure produces intervals containing p 95 percent of the time. In that sense, you have 95 percent confidence that I contains p. If we want more confidence, 98 percent, say, then we must consult the standard normal tables and find an analog to formula 15.2, which is

$$P(-2.33 < W < 2.33) \approx .98$$

from which a 98 percent confidence interval is constructed just as above in the 95 percent case. This interval is obtained by replacing 1.96 by 2.33 in formula 15.3. For given values of n and the sample mean, the 98 percent confidence interval is larger than the 95 percent one. This is to be expected; more confidence has to be paid for by the loss of ability to localize p (unless you take more observations and increase the value of n). The extreme case of this is the 100 percent confidence interval—this is the entire line, and offers no information whatsoever about the location of p.

Let's return to the discussion of the 95 percent confidence interval. The quantity

$$D = 1.96 \sqrt{\frac{\overline{X}(1 - \overline{X})}{n}} \qquad (15.4)$$

defines the interval by giving the deviation from the estimator $\overline{X}$. Is there a way to control how large D can get? An easy calculus argument shows that the product $x(1 - x)$ for $0 \le x \le 1$ has a maximum value when $x = .5$ and the maximum value is .25. Apply this result to the product $\overline{X}(1 - \overline{X})$, using the fact that the proportion $\overline{X}$ is between 0 and 1, and replace 1.96 by 2 in formula 15.4 to conclude

$$D \le 2\,(.5)\sqrt{\frac{1}{n}} = \sqrt{\frac{1}{n}}. \qquad (15.5)$$

This shows that the confidence interval defined by D can be made as small as desired provided only that a sufficiently large number of observations be made. If $n = 1154$, as in the *New York Times* article, from formula 15.5 we see that the deviation from the sample mean is at most $1/\sqrt{1154} \approx .03$. Now let's go back to the quotation from the article given at the beginning of this section. The phrasing " ... in 19 cases out of 20 the results based on such samples will differ by no more than three percentage points ..." indicates a 95 percent confidence interval, where the upper bound .03 can be taken for D because of the sample size 1154. If this upper bound is used with the estimate .56 for the sample mean, we get the confidence interval (.53, .59).

It's interesting to observe the connection between confidence intervals and hypothesis testing. The confidence interval just obtained can be used to test the hypothesis $H_0 : p = .56$ against the alternative $H_1 : p \neq .56$. It works like this: if the null hypothesis is true, $\overline{X}$ should fall outside the confidence interval only 5 percent of the time, so any reading outside of the confidence interval will have p-value smaller than .05. This should lead to rejection of H_0 if .05 is deemed a small enough level to warrant rejection. If a smaller level is desired, find a confidence interval with larger confidence coefficient.

15.7   Random sampling

The mathematical techniques of the statistician require that random samples be taken from the population under study. You can see from the preceding sections that underlying our basic ideas were the two great theorems of probability, the Law of Large Numbers and the Central Limit Theorem, and both of these are about *independent and identically distributed* random variables. This means that each observation is supposed to be a random variable with the same distribution as the basic variable under study, and the observations are independent. So, as we saw in Chapter 12, if I am studying the height, X, of adult American males, I should not expect to get a reasonable estimate of this quantity by selecting observations Y of males living, say, on the west coast. The normal distribution of X describing the entire population may look very different from the normal distribution of Y describing the subpopulation. What I must essentially do is consider the entire adult American male population as items on a huge list; the ith observation X_i will be the result of choosing one of these items from the list at random, that is, using a uniform distribution. We have already seen (in Chapter 13) how to use a random number table to make a random selection from a population. The procedure described there works for a relatively small homogeneous population; for extremely large and more complex populations, more sophisticated techniques are necessary. A form of *stratified sampling* is used, for instance, when the population includes all American adult males. Under this procedure, the United States is broken up into regions, and within each region areas of similar population sizes are grouped. Then a random sample of these areas is selected. Once the area is chosen, there are further stratifications and random choices down the line. For example, in a given area we may want to choose individuals from various geographical parts of the area or from a variety of socioeconomic backgrounds. Once the sample individuals are actually selected, another problem surfaces—how to actually collect the data. A study may be seriously biased, for example, by ignoring those individuals who do not respond for a variety of reasons, perhaps because they are not at home when the data collector arrives. There are statistical devices for eliminating bias that may enter due to such situations. Random sampling, you can see, must be a carefully thought out procedure to make sure the data we collect can validly be used in our statistical theory.

A famous case of bad sampling is the *Literary Digest* magazine poll in 1936 in which the *Digest* predicted an easy victory for Alf Landon over Franklin D. Roosevelt in the presidential election. It turned out that Roosevelt won a landslide victory over Landon. The *Digest*, it appears, had selected its sample from telephone books and club membership lists which biased the sample in favor of the rich. Because it was a Depression year, and the vote of the poor was mostly for Roosevelt, the bias in the sample led the *Digest* to its error. In 1948, a similar instance of bad sampling took place when three major polls predicted the victory of Thomas Dewey over Harry Truman in the presidential election. Again, the sample was biased in favor of the rich as well as having other problems. A lot of people thought Truman didn't have a chance on account of these faulty polls. There is a photograph of a victorious Truman, grinning broadly and holding aloft a newspaper with a banner headline proclaiming his defeat. The theory of sampling had to learn from a lot of bad experience.

15.8   Some concluding remarks

In this chapter, you have seen a few examples illustrating three basic statistical activities: hypothesis testing, and point and interval estimation. I have tried to get across the flavor of some of the standard classical statistical ideas. Many important names are connected with this theory—some pioneers were J. Neyman, E.S. Pearson, and "Student," the statistician W.S. Gossett, who used a pseudonym because he was an employee of the Guinness Brewery (the brewery did not want to advertise to its competitors that the use of statistical reasoning could actually lead to a better product). Before ending this chapter let me briefly mention a few other important approaches to statistical theory.

Much of what we did in this chapter was based on finding an approximately normal distribution for a rather large sample, and much of the classical theory was developed for normal or approximately normal distributions. More recently, the theory of non-parametric statistics has been quickly developing. This subject deals with distributions whose basic form need not be normal; its results remain true whatever the distribution, so the methods are called distribution-free or non-parametric. Non-parametric theory is well developed and powerful.

Another idea was given an impetus during the Second World War. A refinement of the classical theory of hypothesis testing was developed at this time by the statistician Abraham Wald as a direct result of the attempt to save time and money during wartime. In the classical theory, the sample size n is decided upon in advance. Then n observations are taken and a decision is made to accept or reject H_0 based on this information. The method of *sequential analysis* given by Wald does not decide on a fixed sample size in advance. The experimenter takes observations and after each observation a decision is made based on the observations up to that point either to accept or reject H_0 or else *to take another observation*. In this method n, the sample size, is not a constant but a random variable.

The situation can be modelled by a gambler's ruin set-up (Chapter 10). After each play, the gambler is in one of three states: either he is ruined and the game is over, or he ruins his opponent and the game is over, or he plays again. The experimenter using sequential analysis is also in one of three states after each observation: either H_0 is accepted, or it is rejected, or another observation is taken. So the experimenter can be thought of as the gambler, an observation as a play of the game, and acceptance or rejection of H_0 as either gambler's ruin or the opponent's ruin. Continuing the game is equivalent to postponing the decision about H_0 until more observations are taken. Using the methods and ways of thinking in the gambler's ruin model allows the statistician to calculate important quantities, such as the expected number of observations needed. The importance of the method of sequential analysis is that under appropriate conditions sequential tests are more cost efficient than classical tests: you can get the same accuracy as the classical tests with considerably fewer observations on the average.

A viewpoint of statistics philosophically very different from the classical school is the Bayesian approach. The Bayesians take their name and starting point from Bayes's formula (Chapter 4). You recall that Bayes's formula could be interpreted as a method whereby an initial probability $P(A)$ of an event A could be updated by means of additional information B to give the conditional probability $P(A/B)$. Now the classical non-Bayesian approach to the statistical problem involving, say, the probability p of a coin falling head is that p is some *constant* which is unknown to the statistician. On the other hand, the Bayesians regard p as a *random variable* with some given *prior* distribution [corresponding to $P(A)$ in Bayes's theorem]. The information in the sample is used to obtain what is called a *posterior* distribution for p. The posterior distribution is essentially given by the conditional probability $P(A/B)$ in Bayes's theorem. Bayesian statisticians then work with the posterior distribution to draw inferences. There are many problems for which the Bayesian approach is more satisfying than the non-Bayesian one, but the non-Bayesians often counter that the use of subjective probability in determining priors is invalid. Their perspective is that parameters should be unknown constants, not random variables, and the data at the time of the experiment should be all, with any prior information built into the statement of the problem. Most likely, an optimal theory should have elements from both points of view.

There can be almost a religious fervor in the split between the two camps, so it is reassuring to note that the results of both Bayesians and non-Bayesians generally complement each other rather than lead to fundamentally different answers. Perhaps the strength of the emotions churned up is displayed most tellingly by the Bayesian wit who claimed that non-Bayesians may have a rear end, but, as everybody knows, they have no posterior!

15.9   Exercises for Chapter 15

1. Consider 100 Bernoulli trials with success probability p. Using the method of maximum likelihood, describe a test of the hypothesis $H_0 : p = 1/3$ against the alternative $H_1 : p = 2/3$.

2. Suppose a binomial probability p could only be one of N fixed possible values. Describe a procedure depending on the method of maximum likelihood for deciding which of the competing values to accept as the true value of p.

3. A pair of dice is rolled 6000 times and the frequency of 7 is observed to be 900. Let p be the probability that these dice roll 7 on any trial, and let $H_0 : p = 1/6$ and $H_1 : p \neq 1/6$. Use the normal approximation to the binomial distribution to test H_0 against H_1. (Hint: the

percentages of total area lying within 1, 2, and 3 standard deviations of the standard normal distribution are approximately 68, 95, and 99.8, respectively.)

4. Groucho is running for the post of King of Fredonia. His advisers sample 1000 voters and discover that 460 plan to vote for Groucho. Find a 95 percent confidence interval for the proportion of all voters voting for Groucho. Are the results encouraging for the election? Now do this exercise in case 480 of the 1000 voters plan to vote for him.

5. Eight-hundred fish are caught in a lake, tagged, and then released back into the lake. After a while 400 fish are caught and 250 of them are found tagged. Estimate the number of fish in the lake. What are your assumptions?

6. In 6000 rolls of a pair of dice, the frequencies of 2, 3, 7, 11, and 12 are 170, 360, 1150, 340, and 160, respectively. Give point estimates for the probabilities that a single roll with these dice produce (a) 7, (b) 7 or 11, (c) 2, 3, or 12, (d) 2 or 12, (e) none of the five numbers.

16

Roaming the Number Line with a Markov Chain: Dependence

> Time present and time past
> Are both perhaps present in time future,
> And time future contained in time past.
> T.S. Eliot, *Burnt Norton*

16.1   A picnic in Alphaville?

In Alphaville, it has been observed that whether a day is wet or dry depends only on the knowledge of whether the preceding day was wet or dry. Data accumulated over a period of years indicates transitions between wet and dry days are approximately in accordance with the following probability description:

	TOMORROW	
TODAY	DRY	WET
DRY	.6	.4
WET	.2	.8

The table indicates, for instance, that the conditional probability of a dry day tomorrow given that today was dry is .6, the conditional probability of a dry day tomorow given that today was wet is .2, etc. Suppose we are planning a picnic for Sunday and today is Friday. What is the probability that Sunday will be dry given that it is dry today?

We are going to solve this problem using the conditional probability ideas introduced in Chapter 3. Now, however, we want to look at the problem from a somewhat different point of view involving random variables. Let us suppose that for each non-negative integer n we have a random variable X_n which is either 0 or 1. We interpret X_n as the state of the weather (0 for a dry day, 1 for a wet day) on the nth day after the initial day 0. In our problem, day 0 is the particular dry Friday before the picnic. Since Sunday is the second day after day 0, in terms of the random variables X we are being asked to calculate $P(X_2 = 0/X_0 = 0)$. Let us consider the *paths* from Saturday to Sunday; this is simply the list of all possible values of the variables X_1 and X_2. There are four such paths: (0,0), (0,1), (1,0), (1,1), where the first coordinate in the pair denotes the value of X_1 and the second the value of X_2. In the following descriptions, we are going to express a set like $\{X_1 = 0 \text{ and } X_2 = 0\}$ as $\{X_1 = 0, X_2 = 0\}$; that is, for notational convenience we will replace the word "and" by a comma. Note that the conditional probability of (0,0) given $X_0 = 0$ can be written

$$P(X_1 = 0, X_2 = 0/X_0 = 0) = \tag{16.1}$$
$$P(X_1 = 0/X_0 = 0) \, P(X_2 = 0/X_0 = 0, X_1 = 0).$$

To check formula 16.1 use the definition of conditional probability in terms of intersection (Chapter 3) to express the left-hand side as

$$\frac{P(X_0 = 0, X_1 = 0, X_2 = 0)}{P(X_0 = 0)}. \tag{16.2}$$

On the other hand, each term on the right-hand side can be expressed in the same fashion to get

$$\frac{P(X_0 = 0, X_1 = 0)}{P(X_0 = 0)} \; \frac{P(X_0 = 0, X_1 = 0, X_2 = 0)}{P(X_0 = 0, X_1 = 0)}$$

which, after cancellation, reduces to formula 16.2, proving the two sides equal in formula 16.1. Now let's look at the second term on the right of formula 16.1, that is,

$$P(X_2 = 0/X_0 = 0, X_1 = 0). \tag{16.3}$$

In words, this can be expressed as the conditional probability that Sunday is dry, given that Friday and Saturday are both dry. But according to the description of the problem, only Saturday's weather affects the weather on Sunday; Friday's weather is independent of it. Another way to say this is that

$$P(X_2 = 0/X_0 = 0, X_1 = 0) = P(X_2 = 0/X_1 = 0). \tag{16.4}$$

The right-hand side of formula 16.4 is .6 from the table since it does not matter which day is actually today: the table gives the conditional probability of the next day's weather, given today's weather. From formula 16.4 we can simplify formula 16.1 to get

$$P(X_1 = 0, X_2 = 0/X_0 = 0) = \tag{16.5}$$
$$P(X_1 = 0/X_0 = 0) \, P(X_2 = 0/X_1 = 0).$$

Similarly, by considering paths such that $X_1 = 1$, the same reasoning shows

$$P(X_1 = 1, X_2 = 0/X_0 = 0) = \tag{16.6}$$
$$P(X_1 = 1/X_0 = 0) \, P(X_2 = 0/X_1 = 1).$$

Now add up the left-hand sides of formulas 16.5 and 16.6; these are conditional probabilities of disjoint events given $X_0 = 0$, so the sum of the probabilities is the conditional probability of the union set given $X_0 = 0$, namely,

$$P(X_2 = 0/X_0 = 0),$$

exactly what we're looking for. This must equal the sum of the right-hand sides of formulas 16.5 and 16.6 which can be calculated by the table to be $(.6)(.6) + (.4)(.2) = .44$. The odds for a dry picnic day are a little less than even. We have solved the problem, but let's examine the structure of the random variables X_n a little further.

From our description, it should be clear that the value of any X_n, $n > 0$, depends only on the value of X_{n-1}, the immediate predecessor of X_n. Up to now we have mostly talked about independent random variables. The variables X_n are not independent and not surprisingly we call them *dependent*. But the X_n have a relatively mild form of dependence: instead of X_n depending on the whole past history of the process, that is, on the values of each X_i for $i \leq n-1$, X_n depends only on the value of X_{n-1}. A sequence of random variables X_n such that the conditional distribution of each variable given the past only depends upon the value of the immediately preceding variable is called a *Markov chain* or *Markov process*. The value of X_n is frequently thought of as the *state* of a moving particle at *time* n. The intuitive idea is that the evolution of the Markov chain describes the position of the particle at unit time intervals. The word *chain* is used primarily when the state space (the set of all states) is discrete. The Markov chain we have been dealing with in the problem above has another very pleasant property: it is homogeneous in time, which means probabilities like

$$P(X_n = 0/X_{n-1} = 0)$$

do not depend on the time n; they only depend on yesterday's state (in this case 0) and today's state (in this case also 0) with only one day's difference between the days. Similarly, a probability like

$$P(X_{n+k} = 0/X_{n-1} = 0)$$

will not depend on the value of n for a homogeneous process, but only depends upon the $k+1$ unit difference between the future time and the given time. All the Markov chains considered in this chapter will be homogeneous, so we will assume this property in the following without further mention.

We can think of a Markov chain as taking values in a *state space*; for us this space will be discrete and can be represented by some subset of integers. In the problem just considered, the state space consists of the integers 0 and 1. The *one-step transition probability* from state u to state v is defined by

$$p_{u,v} = P(X_n = v / X_{n-1} = u)$$

and by a generalization of the arguments given above, we can see without too much trouble that if $s_0, s_1, \cdots, s_n$ is a sequence of states, then

$$P(X_1 = s_1, \cdots, X_n = s_n / X_0 = s_0) = p_{s_0, s_1} \, p_{s_1, s_2} \cdots p_{s_{n-1}, s_n}.$$

This is the conditional probability of the chain traversing the path

$$(s_1, s_2, \cdots, s_n)$$

given that it starts at s_0. The Markov chain can therefore be thought of as starting out at some state at time 0; this is the value of X_0. From there it jumps to another state, and this is the value of X_1, and so forth.

One of the important questions about Markov chains is to describe the long-term behavior of the system. To be more precise, suppose we have a given distribution for X_0, that is, we know what the probabilities are for finding ourselves in each state at the initial time 0. After the chain jumps once, the new positions are described by the distribution of X_1, and so on; after n jumps, the distribution of X_n describes our position. Of great interest is what happens to the distributions of X_n as n gets very large; in particular, will these distributions converge in some sense to a fixed distribution? If this happens, we have a long-term settling down of the process to a stable or stationary distribution. Under rather broad conditions, it can be proved that a stationary distribution exists no matter what the initial distribution of X_0 is. A similar problem is to look at a fixed state s and study what happens to the probabilities $P(X_n = s / X_0 = s)$ when n gets very large—is there also settling down for a single state? Again, under appropriate conditions, there will be. We will come back to these issues later, but now we want to look at another question and solve it for an important class of Markov chains. This problem can be phrased as follows: given that $X_0 = s$, what is the probability of the Markov chain X ever returning to s? We are going to solve this problem for one-dimensional random walks, introduced in Chapter 10 in connection with the gambler's ruin problem.

16.2   One-dimensional random walks

A very important example of a Markov chain is obtained by considering the partial sums

$$S_n = X_1 + X_2 + \cdots + X_n,$$

where the X variables are independent and identically distributed. That S_n is a Markov chain is pretty easy to see intuitively. Let's assume the X's are discrete. You want to see that

$$P(S_n = s_n / S_1 = s_1, S_2 = s_2, \cdots, S_{n-1} = s_{n-1}) \qquad (16.7)$$

only depends upon s_{n-1} and s_n; that is, the probability does not depend on any of the values s_i for $i \leq n - 2$. Formula 16.7 is equivalent to

$$P(X_n = s_n - s_{n-1} / S_1 = s_1, S_2 = s_2, \cdots, S_{n-1} = s_{n-1}) \qquad (16.8)$$

because $S_n - S_{n-1} = X_n$ by definition. Now the given information in terms of S_1 through S_{n-1} just depends on the values of X_1 through X_{n-1}; since X_n is *independent* of X_1 through X_{n-1}, the given part of formula 16.8 isn't really providing new information about X_n. So the probability in formula 16.8 is simply $P(X_n = s_n - s_{n-1})$, which clearly does not depend on the states s for $i \leq n - 2$, and the S_n is a Markov chain as claimed.

When the X variables take integer values, the Markov chain S_n is called a random walk; it is one-dimensional because its states are integers on the line. If we are more specific and require that the X variables only take the values 1 or -1 with probabilities p and q, we get what is called *Bernoulli* random walk; this is the kind of random walk we considered in Section 10.1. We think of S_0 as describing the position of a particle at some given initial state, some integer the random walk starts out from. Then the particle jumps to $S_1 = S_0 + X_1$, which means it jumped to one of the adjacent integers, so that S_1 is one more or one less than S_0. The particle continues jumping in this way as the Markov chain S evolves in time. As n increases, after n steps the particle can be in an ever larger number of possible positions given by the possible values of S_n. The transition probabilities of the chain are given by $p_{u,u+1} = p$, $p_{u,u-1} = q$, and $p_{u,v} = 0$ if v is different from $u + 1$ and $u - 1$.

The Bernoulli random walk is unrestricted: starting from any state there will eventually be a path leading as far away from the state as you might wish. By altering this random walk in a simple way, we will get the random walk used to model the gambler's ruin problem. We set $S_0 = i$ with probability 1; this represents the gambler's initial fortune. Let $a > i$ be the total fortune of the gambler and her adversary. Define the transition probabilities of the *gambler's ruin* chain as follows: if $0 < i < a$, then $p_{u,u+1} = p$, $p_{u,u-1} = q$, just as with the Bernoulli random walk. But then we put $p_{0,0} = p_{a,a} = 1$; this means that once the chain (or the particle) enters either state 0 or state a it cannot leave—there is probability 1 of it jumping to itself. Such states are called *absorbing states*—in the gambler's ruin chain the two absorbing states at 0 and a express the end of the game with either the gambler or her opponent being ruined. The chain has the finite state space of the integers from 0 to a. In Chapter 10, we learned quite a bit about the long-term behavior of the gambler's ruin chain S_n.

We proved that with probability 1 the chain will end up at one of the absorbing states 0 or a, and we found the probabilities of ending up at each. The probability of the gambler's ruin given in formulas 10.6 and 10.8 can be described in the terminology of the gambler's ruin Markov chain S_n as

$$P(S_n = 0 \text{ for some } n > 0/S_0 = i). \tag{16.9}$$

16.3   The probability of ever returning "home"

To give a little flavor of the subject, we are going to study some problems about recurrence for the Bernoulli random walk. Suppose we start off with $S_0 = 0$ with probability 1; that is, the particle whose movements are governed by the chain is initially placed at 0, which we can think of as "home." There is nothing intrinsically special about 0 here; any other state would do as well. What we would like is to find the probability of the particle ever returning to 0. Recall that for a Bernoulli random walk there are paths getting as far away from the starting point as you might wish provided you wait long enough. How many of these paths escape forever by never coming back to the starting point? We will give an answer to this question by using what we know about the gambler's ruin chain as a tool.

Consider the Bernoulli random walk with $p = q = .5$. For this random walk, we can write

$$P(S_n = 0 \text{ for some } n > 0 \ / S_0 = 0)$$
$$= .5 \ P(S_n = 0 \text{ for some } n > 1 \ / S_1 = -1)$$
$$+ .5 \ P(S_n = 0 \text{ for some } n > 1 \ / S_1 = 1). \tag{16.10}$$

Formula 16.10 says something intuitively obvious, but you may not realize it at first just by looking at the formula. The probability of ever returning to 0, given that you start from 0 (left-hand side of the equation), is simply the probability of jumping in the first step to -1 and then returning to 0 at some time in the future from -1, or jumping to 1 in the first step and then returning to 0 at some time in the future from 1. The equality of the two sides of formula 16.10 can of course be proved rigorously by using the conditional probability formulas and the Markov chain property. (The Markov chain property implies that the event $\{S_n = 0$ for some $n > 1\}$, which depends upon the future of the chain after time 1, is independent of the past, given the present at time 1—this fact is needed to prove formula 16.10.) Now let's focus on

$$P(S_n = 0 \text{ for some } n > 1 \ / S_1 = 1). \tag{16.11}$$

Because of the homogeneity property in time, formula 16.11 is the same as

$$P(S_n = 0 \text{ for some } n > 0 \ / S_0 = 1), \tag{16.12}$$

so formula 16.12 gives the conditional probability of the Bernoulli random walk ever reaching 0, given that it starts at 1. We are going to evaluate this probability by means of a little trick. Consider an integer $a > 1$, and consider the gambler's ruin chain with $p = q = .5$. The essence of the trick is to realize that *the Bernoulli Markov chain starting at 1 has the same transition probabilities as the gambler's ruin chain with total fortune a starting at 1 until the Bernoulli chain hits either 0 or a for the first time.* Because of this we can make certain correspondences about probabilities relative to the two chains. For example,

P(the Bernoulli walk hits 0 sometime before hitting $a/S_0 = 1$)

$= P$(the gambler's ruin walk ends in ruin (hits 0 before a)/$S_0 = 1$)

$$= 1 - \frac{1}{a}, \tag{16.13}$$

where the extreme right side of formula 16.13 is given by formula 10.8 with $i = 1$. Now the left-hand side of formula 16.13 always converges to the probability of formula 16.12 as a tends to infinity, that is, as a gets larger and larger. To convince yourself of this, just think of any path from 1 that returns to 0. It must have hit some finite maximal state before the return. If a is taken sufficiently large, this maximal state of the path is smaller than a, and the path will therefore be counted in calculating the left-hand side of formula 16.13. So as a tends to infinity, all such paths will eventually be counted on the left of formula 16.13, and these paths are precisely the ones that have to be counted in calculating the probability in formula 16.12. On the other hand, as a tends to infinity, the right-hand side of formula 16.13 converges to 1, so that we have shown the probability in formulas 16.11 or 16.12 is equal to 1. Because the Bernoulli random walk with $p = q = .5$ is symmetric, we can conclude that replacing $S_0 = 1$ with $S_0 = -1$ in formula 16.11 also must give a value of 1, and then from formula 16.10 we see that the left-hand side is 1. So using a comparison to a gambler's ruin chain, we have proved that *the Bernoulli random walk with $p = q = .5$ starting off at 0 will return to 0 with probability 1.*

Now let's see if we can use the line of reasoning above to figure out what happens for general values of p and q. Assume $p \neq q$, so we have all possibilities except for the one just studied. This general Bernoulli walk gives rise to

$P(S_n = 0$ for some $n > 0$ /$S_0 = 0)$

$= q\,P(S_n = 0$ for some $n > 1$ /$S_1 = -1)$

$+ p\,P(S_n = 0$ for some $n > 1$ /$S_1 = 1)$, $\tag{16.14}$

the analog of formula 16.10. Once again, we concentrate on formula 16.12, and evaluate it by comparing it to a gambler's ruin walk with the same probabilities p and q as the Bernoulli walk. This gives the same first equality of formula 16.13. Since $p \neq q$, the extreme right-hand side of this formula must now be replaced by

$$\frac{\frac{q}{p} - \left(\frac{q}{p}\right)^a}{1 - \left(\frac{q}{p}\right)^a}, \tag{16.15}$$

which comes from formula 10.6 with $i = 1$. Again, we want to let a tend to infinity to see what happens. At this point, formula 16.15 has to be handled carefully. If $p > q$, the fraction $(q/p)^a$ converges to 0 as a gets large, and the expression of formula 16.15 converges to q/p. Using the same reasoning as before, formula 16.13 now allows us to conclude that the value of formula 16.12 is q/p. In other words, if the Bernoulli walk has a drift to the right (i.e., $p > q$), then, starting at 1 there is positive probability $1 - q/p$ of never hitting 0. This result makes intuitive sense since the larger the value of p, the more likely it should be to escape from 0. But what happens if the drift is to the left, that is, $q > p$? In this case, the expression of formula 16.15 "explodes" as a tends to infinity, namely, the fraction $(q/p)^a$ gets larger and larger (i.e., tends to infinity), and the expression can't be evaluated. What to do? Since we want to get back to the nice situation of the larger number in the fraction on the bottom so the ath power of the fraction tends to 0 as before, we rely on the following device: interchange the roles of gambler and adversary. To do this, we calculate the probability of the *adversary's* ruin. To put the adversary in the boots of the gambler simply means that we must change our point of view (if we have been imagining the gambler to be a stand-in for ourselves, we now use the adversary as that stand-in). How do we then calculate the probability of the adversary's ruin when the gambler has \$1? Well, we go back to formula 10.6. Since the adversary is now the gambler of the formula, to use the formula we must interchange p and q in the formula and put in $(a - 1)$ for i. This gives us

$$w = \frac{\left(\frac{p}{q}\right)^{a-1} - \left(\frac{p}{q}\right)^a}{1 - \left(\frac{p}{q}\right)^a}. \tag{16.16}$$

This is the probability of the adversary's ruin when the gambler has \$1, so that $1 - w$ is the probability of the gambler's ruin when she has \$1. If we use a little algebra (a constant refrain in this book), formula 16.16 shows

$$1 - w = \frac{1 - \left(\frac{p}{q}\right)^{a-1}}{1 - \left(\frac{p}{q}\right)^a}. \tag{16.17}$$

What we want is the limiting probability of the gambler's ruin when the gambler has \$1 and a is tending to infinity. Formula 16.15, as we saw, exploded as a got larger and larger. The beauty of formula 16.17 is that we can easily see what happens as a gets larger and larger. Since $p/q < 1$, the second terms on the top and bottom of the ratio in formula 16.17 tend to 0 as a tends to infinity, and so $1 - w$ converges to 1.

The same reasoning as in the case $p > q$ (that is, go back to formulas 16.13 and 16.12 once more) proves formula 16.12 equal to 1. So if the Bernoulli walk has a drift to the left, then starting at 1 there is probability 1 of eventually hitting 0; that is, the probability of escape from 0 is 0. To finish off the problem, formula 16.14 must be evaluated. There are two cases, $p > q$ and $q > p$. In the case where $p > q$, the second term on the right-hand side of formula 16.14 is $p\,(q/p) = q$. To evaluate the first term on the right-hand side in this case, use symmetry as follows: the chain starting from -1 has the same probability of return to 0 as a chain *starting from 1* with transition probabilities q of moving one step to the *right* and p of moving one step to the *left*. We have just seen that this probability is 1. So formula 16.14 evaluates to $q + q = 2q$ in the case $p > q$. Turning to the case $q > p$, now the first term of formula 16.14 by symmetry is equal to $q\,(p/q) = p$ and the second term is equal to p, so in this case formula 16.14 evaluates to $2p$. All of the above results can be put together to give the following statement.

Let a Bernoulli random walk have probabilities p and q of moving one unit to the right and one unit to the left. Then the conditional probability of ever returning to 0 given that you start at 0 is equal to $2m$, where m is the minimum of p and q. This probability is 1 when $p = q = .5$; otherwise there is a positive probability $1 - 2m$ of escape from 0.

In the case $p = q = .5$, the particle returns to 0 with certainty, and it can be shown that the particle returns to 0 *infinitely often* with certainty. Yet it turns out that as n increases, the probability $P(S_n = 0/S_0 = 0)$ of the particle being in state 0 at time n given that the particle started at state 0 converges to 0. So although the particle is certain to return to 0 at some time, the probability that the particle will be there at any fixed large time is very small. This is not too surprising when you think about it: the particle starting from 0 spreads out and eventually is able to visit any state no matter how far away from 0 that state is. So for large n the number of states that could be occupied is very large and the probability of being in any one of these is small. The particle is thus spreading itself thinner and thinner over the state space as time evolves. The probability of the particle being at 0 (or at any other state, for that matter) at any fixed time n therefore converges to 0.

16.4 About the gambler recouping her losses

The results of the preceding section have interesting consequences for the theory of gambling. A Bernoulli random walk can model the non-realistic situation of a never-ending game involving a casino with infinite capital which allows a gambler to amass as much debt as she incurs. We assume the usual situation for the gambler: she is playing an unfavorable game ($p < q$). She starts with \$1, say, and the play continues even if she falls into debt. Suppose she starts to lose (the Bernoulli walk is at some negative integer at some time). What can we say about the gambler recouping her losses?

Suppose the gambler is \$1 in debt; that is, the Markov chain S_k representing her fortune at time k is sitting at -1. The gambler will recoup her losses if $S_n = 1$ at some time after k. Because of the time homogeneity of the Markov chain, we might as well think of $S_0 = -1$ (the gambler is \$1 in debt at time 0) and the gambler recouping her losses at some time $n > 0$, that is, $S_n = 1$. The Markov chain can only get from -1 to 1 by getting to 0 at some time and then at some later time getting to 1 from 0. It follows from the preceding section that the probability of the gambler getting to 0 at some time, given that she starts at -1, is p/q. Once at 0, she gets to 1 eventually with another probability p/q. What happens after she gets to 0 for the first time is *independent* of what happened before she hit 0. It will follow that her probability of ever recouping her losses, given that she is \$1 in debt, is $(p/q)^2$. In similar fashion, we can see that the probability of the gambler recouping her losses if she is \$$k$ in debt is $(p/q)^k$, which shows that the gambler's chance of ever recouping her losses decreases exponentially to 0 according to the power of her debt. The deeper in debt the gambler finds herself, the more unlikely it will be that she will ever recoup.

The above description is for a gambler playing a game unfavorable to her. What happens if she plays a fair game? Consider the Bernoulli random walk for the case $p = q = .5$. The walk is certain to return to any starting point (as we know from the preceding section). We want to make a useful observation for this random walk, namely, with certainty it will attain any fixed state, given that it starts from any fixed state. To see this, it is sufficient (because of symmetry) to show that you can get anywhere with probability 1 from 0. So let a be any fixed state. There is certainly a path (with positive probability) reaching a from 0. By exercise 3 at the end of the chapter, the chain must therefore attain 0 with probability 1, starting from a. By symmetry, we can shift and conclude that the chain must therefore attain the state $-a$ with probability 1, starting from 0. Since $-a$ can be any state, we have proved what we want.

Now let's assume, as before, that the random walk models the gambler's game, which is fair this time. As before, we suppose the gambler starts with \$1 and plays until she is \$1 in debt. But now the random walk sitting at -1 is certain to get back to 1, which in gambling terminology says the gambler is certain to recoup her losses. Let us ask ourselves an interesting question: how long will it take her, on the average, to do this? To see what we are being asked to calculate here, we look at all paths $(-1, s_1, \cdots, 1)$ beginning with -1 and stopping at the first time the walk gets to 1. Let N be the time of first entry into 1; that is, for each such path we have just described, N is one less than the number of terms in the path. Now, N is a random variable and we can take its expectation (for each positive integer k, compute the probability that $N = k$, multiply by k, and then add up over all possible k). It can be shown (but we won't do it here) that N has infinite expectation for the game we are discussing now. What we mean by this is that the sum defined by the expectation does not converge; it gets arbitrarily large as more and more terms are added up. So although the gambler is certain to recoup her losses, it will take her infinitely long, on average, to do so! The practical consequences of this statement mean the gambler can expect a lot of plays before losses can be recouped and the likelihood of sinking more deeply into debt before recouping. So even if the gambler is playing a fair game, if she is in debt she should have a lot of patience and expect to be playing a long time before breaking even.

The discussion above makes the following surprising fact plausible: suppose the gambler is playing a *fair* game of the Bernoulli type. *It is much more likely for the gambler to stay in a winning state or a losing state for long periods of time than to have frequent swings of luck.* For example, the probability is .2 that the gambler will either keep winning or keep losing for almost 98 percent of the time. We can think of this as the Markov chain's reluctance to cross the origin 0 from one side to the other. This is in line with what we have already seen: that the losing gambler should not hope to recoup her losses quickly and may have to go deeply into debt before the certain recoup takes place. Conversely, if the gambler is lucky enough to be in the lead, she is likely to remain in the lead for awhile. So this last fact is finally a bit of good news for the gambler *if* she is playing a fair game and *if* she finds herself in the lead. Of course, a real gambler is typically playing an unfavorable game. So, as we have seen above, the reluctance of the Markov chain S_n to cross from one side of the origin 0 to the other side now becomes even greater, with the chain spending most of its time on the negative debt side with large probability.

16.5 The dying out of family names

In this section, we look at an important class of Markov chains called *branching processes*. A colorful problem to describe this type of process is the question of the dying out of family names. We consider a father who has U sons, where U is a random variable taking on the values $0, 1, \cdots$ with some probability distribution. In turn, each of his sons can themselves have sons whose number is given by the same probability distribution as that of U. Let the father represent the 0th generation, and define the nth generation as the generation of sons following the $(n-1)$st generation. Let X_n be the size of the nth generation (for instance, the size of the second generation is the number of grandsons of the original father). If X_n is ever

0, then $X_m = 0$ for all $m > n$. If $X_n = i > 0$, then

$$X_{n+1} = U_1 + U_2 + \cdots + U_i.$$

The size of the $(n + 1)$th generation only depends on the size of the nth generation, so X_n is a Markov chain; it has as its state space the set of non-negative integers, and, as with the gambler's ruin chain, 0 is an absorbing state.

We are interested in the event $\{X_n = 0$ for some $n > 0\}$, that is, the event that the line of sons is eventually extinct and the family name ultimately dies out. To exclude unimportant cases, assume $P(U = 0)$ is strictly between 0 and 1. Then it is possible to prove that extinction happens with certainty if $EU \leq 1$, but if $EU > 1$ there is a positive probability of the family name continuing forever. You can find these facts proved in [3], for example.

The branching process model applies to many situations. In physics, it can be used in considering particles that can generate a number of other particles when undergoing bombardment. The physicist may be interested in studying the rate at which the number of particles increases. For geneticists, the model applies to genes; the number of descendants in each generation in which the gene is found is the state of the process. The gene may mutate, or not appear in a descendant for other reasons. Using the theory, the probability of long-term survival of the gene can be estimated.

16.6   The number of parties waiting for a taxi

Another class of important Markov chains is obtained from *queueing processes*, which study problems concerning the wait for service in various situations. The individuals doing the waiting can be customers at a bank, programs in a computer waiting to be executed, patients in a doctor's office, etc.; we refer to these individuals as *customers* no matter what they are. Customers arrive for service and are either serviced immediately or form a waiting line, or queue, for the available server. There are a number of obvious questions of interest in such problems. One of these is: after a sufficient amount of time has passed, is there anything we can say about the length of the queue on average? We might also be interested in the expected waiting time for a customer. Many different queueing models can be set up by further specifications, for example, by giving the distribution of the arrival times of the customers, the number of servers, and the distribution of the serving period for the servers. Let's illustrate with a very simple example.

Assume there is a taxi stand at which taxis arrive at fixed time intervals. If no one is waiting, the taxi leaves immediately; otherwise it picks up the first party waiting and departs (a *party* means any group arriving together who can all fit into one taxi). Between taxi arrivals, parties arrive and form a queue; we assume the number of parties arriving in such a period between taxi arrivals is given by a random variable U with some fixed distribution. Let X_n be the number of parties in the queue in the nth time period, that is, between the departures of the $(n-1)$st and nth taxis. Then X_n is a Markov chain. To see this, suppose we know $X_{n-1} = a$; then $X_n = U + a - 1$ if $a \geq 1$ and $= U$ if $a = 0$. So if we know the value of X_{n-1}, the distribution of X_n can be explicitly calculated in terms of the distribution of U. Knowing the distant past history of the process, that is, the values of the X's for the periods less than $n - 1$, does not give any further information about X_n. The X process is therefore a Markov chain with state space the integers $0, 1, 2, \cdots$. The size of the queue for large n, namely, the value of X_n for large n, turns out, not surprisingly, to depend crucially on EU, the expected number of parties arriving in a time period. Since each taxi can only serve a single party, it is intuitively clear that if $EU > 1$ the queue size is going to get abitrarily large as time goes by. What is not so obvious is that if $EU < 1$ the queue size will approach a steady state as time increases. What this means is that there exists a distribution v on the state space such that

$$\lim_{n \to \infty} P(X_n = a) = v(a)$$

for each non-negative integer a. This stationary or steady-state distribution depends upon the distribution of U. In contrast to the above two cases, the single remaining case in which $EU = 1$ gives a chain that neither gets large without bound nor settles down to a fixed distribution; instead, the chain roams all over the state space, returning to each state infinitely often with probability 1. This is a highly unstable queueing process. If we wait long enough, the size of the queue will eventually be any non-negative integer, and yet, for any fixed state a the probability of the chain being in a will tend to 0 as time increases. This kind of behavior has similarities to the Bernoulli random walk with $p = q = .5$.

16.7   Stationary distributions

In the last section, we mentioned a stationary or steady-state distribution v which sometimes exists for a Markov chain. This distribution has the property that, if the chain starts off with this distribution, at the end of one transition the chain has the same distribution. More precisely, suppose the chain has initial probability v, that is, $P(X_0 = a_i) = v(a_i)$ for each state a_i in the state space. Then v is a stationary distribution if, for all states a_i,

$$v(a_1)P(X_1 = a_i/X_0 = a_1) + v(a_2)P(X_1 = a_i/X_0 = a_2) + \cdots$$
$$= P(X_1 = a_i) = v(a_i). \quad (16.18)$$

You can think of it this way: if a particle is initially located at state a with stationary probability $v(a)$, then after the particle undergoes one transition the probability is still $v(a)$ that the particle is located at a. It then seems reasonable to suppose that if the particle is initially located at a with probability $v(a)$, then after the particle undergoes any finite number of transitions the probability is still $v(a)$ that the particle is located at a. This can be proved rigorously, essentially by observing the truth of formula 16.18 if X_1 and X_0 are replaced by X_{n+1} and X_n, respectively.

Not all Markov chains have stationary distributions (for instance, no random walk does—see the exercises at the end of this chapter). The existence of a stationary distribution for a Markov chain is a very important property of that chain since one can prove a lot of nice things about such chains. How could we go about finding out whether a chain has such a distribution? One way is simply to assume a stationary distribution exists and try to find it by means of formula 16.18. As an example, consider the chain given in Section 16.1, where the state space consists of the two integers 0 and 1, and the one-step transitions are given by the table. If we let

$$p_{i,j} = P(X_1 = j/X_0 = i),$$

and we assume there is a stationary v, then formula 16.18 is equivalent to the two equations

$$v(0) = v(0).6 + v(1).2 \text{ and } v(1) = v(0).4 + v(1).8.$$

Moreover, we know that v is a probability, that is, $v(0) + v(1) = 1$. These three equations have a unique common solution $v(0) = 1/3, v(1) = 2/3$.

16.8   Applications to genetics

We are going to assume a model of random mating. Let's say there are three genotypes in a population, described by the pairs of genes AA, Aa, aa, occurring in the ratio $u : 2v : w$. Two individuals are taken at random from the population and mated. Each individual contributes one of its pair of genes with probability .5 independently of its mate to form the genotype of its offspring in what we can call the first generation. The problem is to find the probability distribution of the genotypes in this first generation.

To do this, it is convenient to think of the individual genes A and a floating around in a container and joining up at random to form a genotype. We can find the ratio of A to a genes as follows. Assume the numbers $u, 2v$, and w give the exact numbers of each genotype pair. If we think of splitting the pair, we would get $2u + 2v$ A genes, and $2w + 2v$ a genes. So the ratio of A to a genes is $u + v : w + v$, and now if we suppose the ratio $u : 2v : w$ is standardized with $u + 2v + w = 1$, then we can put $p = u + v$ as the probability of the A gene in our pool, and $q = w + v$ as the probability of the a gene, where $p + q = 1$. Under random mating, the genes would join independently: the genotype AA occurs when the male and female contribute the A gene; this occurs with probability p^2. The genotype Aa occurs when either the male contributes the A gene and the female the a gene, or the other way around; the probability is $2pq$. The genotype aa occurs when both parents contribute the a gene; this probability is q^2. So now we see the genotypes AA, Aa, and aa in the first generation occur in the ratio $p^2 : 2pq : q^2$. Now let's see what happens in the second generation. Repeating the argument above to determine the ratio of A to a genes from the ratio of the genotypes, we obtain $p_2 = p^2 + pq$ as the probability of the A gene in the second generation and $q_2 = q^2 + pq$ as the probability of the a gene in the second generation. So for the second generation, we have:

$$P(AA) = (p^2 + pq)^2 = (p(p+q))^2 = p^2, \text{ and}$$

$$P(aa) = (q^2 + pq)^2 = (q(q+p))^2 = q^2.$$

These are precisely the same probabilities for AA and for aa for the first generation, and it follows without further computation that the probability for Aa must be the same as for the first generation as well. Therefore, we have shown that the distribution of the genotypes is the same in the first and second generations, and it is clear that the distribution remains the same for all generations $n \geq 1$. In fact, let's see that we have a Markov chain with a stationary distribution. Let $X_n = 0, 1$, or 2 depending on whether a genotype of the nth generation is AA, Aa, or aa, respectively. The initial or 0th generation, X_0, has the initial distribution given by the ratio $u : 2v : w$. The Markov property holds because the distribution of the genotypes at any generation n clearly determines the distribution of the genotypes at the succeeding generation $n+1$, so the distant past (knowledge of distributions before generation n) is irrelevant. But our computations show that under random mating the distributions of X_n are stationary for $n \geq 1$. This statement is known as the Hardy-Weinberg law. The stationarity shows stability of the genotype distribution as time evolves. In practice, the gene frequencies p and q are affected by random fluctuations as the process evolves, so the actual distributions will deviate from the theoretical stationary one as time goes on. In fact, if the population is bounded in size, one can show that eventually one gene should die out, leaving one of the types AA or aa. But this too is a theoretical result based on a simplified model; in reality, mutations and other biological effects make the situation more complex.

16.9 Exercises for Chapter 16

1. A famous billionaire always plays red at roulette, betting $1000 at each game. Suppose he lost $5000 by losing each of the first five games.

Assuming he is allowed to go into debt as much as is necessary and keeps playing, estimate the probability of his ever recouping his losses.

2. Consider the random walk with p and q the probabilities of moving one step to the right and left, respectively. (a) Given that $X_0 = 0$, find the probability that $X_4 = 0$. (b) Given that $X_0 = 0$, show that $X_n = 0$ has probability 0 if n is an odd number.

3. Consider any Markov chain with state space the integers. Suppose it is known that the chain, starting from 0, returns to 0 with probability 1. Let a be a state that can be reached from 0, that is, there exists a time n such that $P(X_n = a/X_0 = 0) > 0$. Frame an intuitive argument showing that, starting at a, the chain must eventually reach 0 with probability 1. Try to make your argument rigorous by translating your intuitions into careful mathematical statements.

4. Can a random walk have a stationary distribution? (Hint: if such a distribution v exists, there must exist a state a with $v(a)$ a maximum value, i.e., for any other state b, $v(b) \leq v(a)$. What does this say about the adjacent neighbors of a?)

5. A Markov chain is defined as follows: the state space is the set of non-negative integers 0,1,2, etc. The one-step movement of the chain can be described in words by saying that a particle starting at any integer moves one integer to the right with probability .5 or else hops to 0 with probability .5. The one-step transition probabilities are given by the relations

$$p_{i,j} = \begin{cases} P(X_{n+1} = j/X_n = i) = .5 & \text{if } j = 0 \text{ or } j = i+1, \\ 0 & \text{otherwise.} \end{cases}$$

Does this chain have a stationary distribution v? If so, find it.

17

The Brownian Motion, and Other Processes in Continuous Time

> They have given an amusing explanation of certainty; for after
> demonstrating that all their paths are certain, they no longer
> describe as certain the one that leads to heaven, without any
> danger of not reaching it, but the one that gets us to heaven
> without any danger of diverging from that path.
>
> Blaise Pascal, *Pensées*

17.1 Processes in continuous time

As we have seen, a sequence of random variables can describe how some
process evolves in time. For example, the gambler's accumulated winnings
at time n, that is, after the nth play of a game of chance, is given by
the random variable $S_n = X_1 + \cdots + X_n$, where the X's are independent
random variables representing the gambler's winnings at each play of the
game. For convenience, we can think of the subscript n of any such sequence
of random variables as denoting the nth moment of time, and the value of
the random variable S_n as the measurement made at that moment. The
sequence therefore represents a "picture" of the evolution of the process at
the times $1, 2, \ldots$.

For many physical processes, however, the natural parameter of interest
is continuous, not discrete, time. Let's represent the measurement at time
t by a random variable X_t for all t on some interval, say $0 \leq t$. If we
watch the evolution of the process in time, we get a fixed curve in a t-x

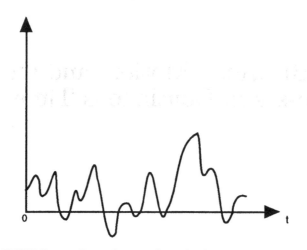

FIGURE 17.1. Part of a sample path of a stochastic process

plane where the observed measurement at time t is plotted at x (see Fig. 17.1). Such a curve is called a *sample path* or *realization* of the process. We have already briefly discussed such processes at the end of Chapter 9, with reference to the Poisson process. When time is discrete, sample paths are just sample sequences, so a typical sample path of the process representing your accumulated winnings if you play a game of chance repeatedly is just the sequence of these winnings on any particular evening. We have been referring to such paths in discrete time throughout this book, but first started using the term *path* when we studied random walk in Chapter 16. In continuous time, a typical path is a curve and the set of all possible curves representing sample paths gives the totality of possible observations for the process evolving in time.

In Chapter 9, X_t was a Poisson process that measured the total number of vehicles passing by a café up until time t, where time 0 is some fixed designated initial time. What would a typical sample path for such a Poisson process look like? Well, let's say that at the moment we start measuring time while sipping a cappuccino at the café there are no vehicles in sight, so the curve starts out at the level 0 and continues there until a first vehicle passes at time t_1. At this point, the curve jumps to level 1 and continues at this level until at time t_2 the next vehicle passes, at which instant the curve jumps to level 2 and continues at that level for awhile, and so forth. Each time a vehicle arrives, the curve jumps one unit higher and continues at that level until the next vehicle arrives. The curve generated in this way is called a step-function (see Fig. 17.2). For any time s that is not an instant in which a vehicle arrives, the value of X_s, namely, the height of the curve above the t-axis at the point s, is precisely the number of vehicles that have passed the café up to the instant s. If s is equal to one of the instants at which a vehicle arrives, there is a little ambiguity about what X_s should

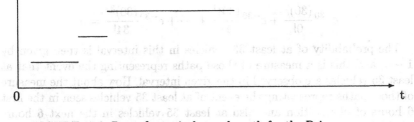

FIGURE 17.2. Part of a typical sample path for the Poisson process

be—should it be the smaller value before the jump took place or the larger value after the jump? This is not a very serious problem; just define the function to be either value at such points. Usually, the value taken is the larger value, and then the sample path is what we call a *right-continuous* function.

What we want to be able to do for a general process in continuous time is find probabilities of events defined in terms of the random variables of the process. For example, consider the Poisson process and the event that there are at least 35 vehicles within the first 6 hours of observation. This is really a statement about the sample paths: we look at all possible sample paths which attain the height 35 and record the first instant this height is attained. If this instant is less than 6 hours (or 360 minutes, or whatever units we are using for measuring time), this path belongs to the event, or set, we are interested in. To discuss the probability of this event, then, we need to have a probability mass on the path space, the set of all possible sample paths. Given an event, namely, a statement about a set of sample paths, we want to be able to compute the probability of this event using the probability mass. For the Poisson process, this probability mass can be constructed from the assumptions (a), (b), and (c) discussed in Chapter 9. For other processes, a variety of assumptions are made which determine an appropriate probability mass. Without such a mass, we technically have no well-defined process and can make no meaningful probability statements.

17.2   A few computations for the Poisson process

For the Poisson process, as we have noted above, the mass (that is, probability) of sample paths satisfying various conditions can be calculated by the methods of Chapter 9. For example, in the problem above we wish to find the mass of the set of paths representing that at least 35 vehicles are seen in the first 6 hours of observation. Let X_t be the number of vehicles observed over the interval from time 0 up until the instant t. It makes sense to suppose no cars have been observed at the initial instant, that is, $X_0 = 0$ with probability 1. Assume the density $\lambda = 5$ as in Chapter 9. According to formula 9.2, the probability of *at most* 34 vehicles passing in a 6-hour interval [i.e., $P(X_6 \leq 34)$] is given by adding up the probabilities that there are 0 vehicles, 1 vehicle, etc., up to 34 vehicles, that is,

$$e^{-30}\frac{(30)^0}{0!} + e^{-30}\frac{(30)^1}{1!} + \cdots + e^{-30}\frac{(30)^{34}}{34!} = u.$$

The probability of at least 35 vehicles in this interval is then given by $1 - u$, and this is a measure of those paths representing the event that at least 35 vehicles are observed in the given interval. How about the measure of those paths representing the event of at least 35 vehicles seen in the first 6 hours of observation and also at least 35 vehicles in the next 6 hours of observation? Assuming the same Poisson process describes the traffic on the road throughout the 12 hour period, we can get the probability by using two of the basic properties of the Poisson process: the independence of the increments of the process over non-overlapping intervals and the dependence of the distribution of Poisson events in an interval only upon the length of the interval, not its endpoints. In terms of the process X_t, we have

$$\begin{aligned} P(X_6 > 35 \text{ and } X_{12} - X_6 > 35) &= P(X_6 > 35)\, P(X_{12} - X_6 > 35) \\ &= P(X_6 > 35)^2 = (1 - u)^2. \end{aligned}$$

17.3   The Brownian motion process

Robert Brown was an English botanist who, in 1827, discovered the random movement of small particles suspended in a fluid. The movement, called Brownian motion, is due to a large number of collisions the particles sustain from the molecules of the fluid. Einstein (and others) worked on a physical theory to explain the Brownian motion. As usual, the physics of the situation used a lot of mathematics in very loose ways and then mathematicians entered the picture to try and make a reasonable mathematical model of the process. Norbert Wiener was the first to arrive at a rigorous description, and in his honor the Brownian motion process is often called the Wiener process.

Let's see what the ideal Brownian particle does. Rather than handle three dimensions, we are going to look at only a single dimension; that is, we are going to see how the particle's x position is changing if its position in three dimensions were plotted in a standard x-y-z coordinate system in space. Since the movements in the three dimensions are independent of each other and identically distributed, studying what's happening in the x direction is all that's necessary to get a complete probabilistic description of the movement of the particle. Now let some initial time $t = 0$ be designated, and let us represent the position X_t of a Brownian particle moving on the x axis. Here are the basic assumptions for the Wiener process, sometimes called the standard Brownian motion:

a. For any $n \geq 3$ times $t_1 < t_2 < \cdots < t_n$, the random variables

$$X_{t_2} - X_{t_1}, \cdots, X_{t_n} - X_{t_{n-1}}$$

are independent (this is often called the assumption of *independent increments*; the Poisson process also has this property).

b. For any $s < t$, the random variable $X_t - X_s$ has a normal distribution with expectation 0 and variance $K(t - s)$, where $K > 0$ is some fixed constant (another way to say this is that the differences, or increments, are normal with mean 0 and variance proportional to the time difference).

Usually the particle is considered as having started at time 0 from the origin, that is,

c. $X_0 = 0$ with probability 1.

Using these conditions, a probability mass, called *Wiener measure*, can be set up on the path space such that the typical Wiener path is continuous (i.e., the points of the curve for t very close to s are also very close), but is extremely "jagged." Mathematically, this means the path does not turn smoothly anywhere—the tangent to the curve, which represents the velocity of a particle moving on the curve and is expressed by what is called in calculus the *derivative* of the function giving the curve, does not exist. We can think of the curve as being very pointy, as though it is always changing direction. This is perhaps not very surprising when you think of the physical Brownian movement with a particle constantly undergoing random bombardment. It is impossible to draw an adequate picture of a Brownian path; the real path is much less smooth than any picture could be.

That a typical Brownian path has tangents nowhere is very interesting in light of the mathematics of the eighteenth and nineteenth centuries. Most mathematicians of that era believed all continuous curves had to have well-defined tangents, and there were repeated attempts to prove this. Finally, in the middle of the nineteenth century, the German mathematician

Karl Weierstrass gave an example of a continuous curve without tangents anywhere. This example must have appeared as an aberration to most of his contemporaries, the type of example mathematicians often call *pathological* to indicate its unusual and unexpected nature. But from the point of view of Brownian paths (which, of course, were unknown at the time) the situation is just the opposite: the typical Brownian path has no tangents, and the totality of paths *with* tangents are the anomalies, having probability 0.

17.4   A few computations for Brownian motion

We can find the Wiener measure of various events by using the assumptions for the Brownian motion process just as we used the assumptions for the Poisson process in the preceding section to calculate probabilities. We are going to consider some easy examples. Let X_t be the position of the Brownian motion at time t. We can ask what the Wiener measure is for the set of paths with $X_4 > 2$, say. According to our assumptions, $X_4 = X_4 - X_0$ is normally distributed with expectation 0 and variance $4K$, where K is the constant for the Brownian motion given in (b) above. So this problem can easily be solved by using the methods of Chapter 12: find the probability that a normal random variable with given mean and variance exceeds a certain value. Now let's compute a conditional probability. Suppose we want to find

$$P(X_4 > 2/X_1 = 3),$$

that is, the probability that the Brownian motion exceeds 2 at time $t = 4$, given that at time $t = 1$ it was at level 3. The above probability is equal to

$$P(X_4 - X_1 > -1/X_1 - X_0 = 3) = P(X_4 - X_1 > -1)$$

by the assumption of independent increments for Brownian motion [property (a) above]. Finally, by property (b), the distribution of the random variable $U = X_4 - X_1$ is normal with mean 0 and variance $3K$, so the methods of Chapter 12 easily lead to a solution of this problem also. From the foregoing, we see something very interesting about Brownian motion. If s is a fixed time and a is any position value with $X_s = a$, the process $U_t = X_t - a$ for $t \geq s$ is a Brownian motion just like X_t for $t \geq 0$. [Another way to say this is that if we relabel the t-x axis of the Brownian motion X_t by shifting so that the point (s, a) becomes the origin of a transformed coordinate system t'-x', then $U_{t'}$ is a Brownian motion with the same parameters as X_t.] It is also intuitively reasonable and indeed true that the U_t process for $t > s$ is independent of the X_t process for $t \leq s$. This follows from the assumption that the Brownian motion has independent increments, as follows. Suppose we want to know $P(U_t > c/X_s = a, X_w = b)$, say, where $w < s$ and $t > s$. This is the same as

$$P(X_t - X_s > c/X_s = a, X_w = b) =$$

$$P(X_t - X_s > c/X_s - X_w = a - b, X_w - X_0 = b) = P(X_t - X_s > c)$$

by the independent increment assumption, so the right-hand side only depends upon X_s and not upon the position of X_w. This essentially shows the Brownian motion process is a Markov process (the future only depends upon the most recent past), about which we'll say more later.

The Wiener measure of more complicated events frequently requires the use of calculus methods. For example, suppose we wished to calculate the Wiener measure of all paths such that $X_1 > 0$ and $X_2 < 0$. A typical path of this type would have $X_1 = a > 0$ and $X_2 = b < 0$ for some numbers a and b. If we had a way to assign an "infinitesimal probability" to each single path of this type, we could then "add up" all these "probabilities" over all possibilities for a and b to get a probability for the union set. Using calculus, such problems can be solved since the Wiener measure defines such infinitesimal probabilities and the ideas of calculus let us handle such objects.

Descriptions of events in continuous time can get much more complicated than in discrete time. Take the event that simply says the Brownian motion stays above the 0 level for all times t between 0 and 1, that is, $X_t > 0$, $0 \leq t \leq 1$. This event depends on *all* times on the unit interval, and this is a continuum of time values. Such events depending on a continuum of times can lead to complications that don't exist in the discrete case, but there are techniques for handling events of this type. This is fortunate, since this kind of event is often of great interest. We give a final computation in this section involving an event defined in terms of a continuum of times but where things turn out simple enough so that we can arrive at an answer easily. We want to consider the event $E =$ "the Brownian particle travels on an increasing path everywhere on the time interval 0 to 1." What this means is that the curve X_t is always increasing on the unit interval. We show that this event has probability 0, that is, the sample paths corresponding to this event have Wiener measure 0, so that the typical Brownian particle is almost surely going to dip down in this interval. To see the truth of this is relatively easy. Let N be any fixed positive integer, and divide the unit interval into subintervals by choosing any instants $0 < t_1 < t_2 < \cdots < t_N < 1$. Then if E is true, it must be the case that each of the conditions

$$X_{t_1} > 0, X_{t_2} - X_{t_1} > 0, \cdots, X_{t_N} - X_{t_{N-1}} > 0$$

are satisfied. But these events are independent, so the probability of their joint occurrence is the product of the probabilities of the separate occurrences. Now take, for example, $X_{t_2} - X_{t_1}$. This is a normally distributed random variable with mean 0, so

$$P(X_{t_2} - X_{t_1} > 0) = .5.$$

Similarly, each of the other random variables to the left of the ">" symbol are normal with mean 0, so each probability is .5. It follows that the event described by the joint occurrence of the N inequalities above is $(.5)^N$. From our laws of probability, we must have $P(E) \leq (.5)^N$ [if F is the event corresponding to the fulfillment of each of the above inequalities, then $E \subset F$, so $P(E) \leq P(F)$]. But the number N of instants we chose was arbitrary, so that $P(E)$, being less than or equal to the Nth power of .5 for any N, no matter how large, must be equal to 0. This proves that the set of Brownian paths which are always increasing on the unit interval has probability 0. Now, there was nothing special in our use of the unit interval here; the argument we just used about choosing N instants could have been made had we been given *any* interval. So what we have really seen is that the Brownian particle can never keep increasing on any interval, and the typical Brownian path is very irregular, always changing direction. This ties in with the property (already mentioned) that the Brownian paths have no tangents anywhere. For if such a path increased everywhere on an interval, the existence of tangents to the path could be shown to exist at (most) points of the path over the interval where the path increases.

17.5 Brownian motion as a limit of random walks

You may have been struck by a similarity of the Brownian motion process to the random walk processes discussed in Chapter 16. In the random walk situation, time is discrete, as is the distance jumped by a particle, but in both cases we have a particle starting out at 0 and performing random movements where the position of the particle at any time, n or t, is given by S_n or X_t. We would like to present an intuitive approach showing how Brownian motion can be derived by thinking of it as a limit of random walks. Our exposition essentially follows the ideas of Kac [20].

Think back to the random walk with $p = q = .5$ (often called the *symmetric random walk*). In this case, a particle starting out at 0 moves a unit to the right or a unit to the left in unit time with probability .5 such that the displacements are independent. What we want to do now is think of the same situation, except that the two units, jump distance and time, will be thought of as very small. Let the unit of time be τ and suppose that at each such unit of time the particle instantaneously moves a distance Δ units from where it is either to the right or to the left with probability .5. More precisely, define the random variable $U_{i\tau}$ to be the displacement at the ith step, where

$$U_{i\tau} = \Delta \text{ or } -\Delta \text{ each with probability .5.}$$

The sequence of displacements $U_{i\tau}$ are assumed independent. Suppose the particle starts off at 0. The position of the particle at time t is roughly given by

$$X_{m\tau} = U_\tau + U_{2\tau} + \cdots + U_{m\tau}, \tag{17.1}$$

where m is the integer part of the quotient t/τ. Formula 17.1 is the sum of independent, identically distributed random variables, and is essentially just another way of writing our old friend

$$S_n = X_1 + X_2 + \cdots + X_n$$

from previous chapters. The variables U satisfy:

$$EU = .5(\Delta) + .5(-\Delta) = 0, \text{ and } \sigma^2(U) = .5(\Delta^2) + .5(\Delta^2) = \Delta^2.$$

It follows from Section 8.8.2 that the variance of $X_{m\tau}$ in formula 17.1 is the sum of the variances of the U random variables, that is, $\Delta^2 m$.

So far τ and Δ have been fixed constants which we have assumed to be small, but now suppose we let both τ and Δ tend to 0, so the time interval between displacements is shrinking to 0 as well as the size of the displacement. In addition, since the time t is a fixed value, the number of displacements m must increase. The trick to get something worthwhile out of all this is to let τ and Δ go to 0 in a special way: it is required that

$$\frac{\Delta^2}{\tau} \to K, \text{ a constant, and } m\tau \to t, \qquad (17.2)$$

where the arrows denote convergence, that is, the left side of the relations get closer and closer to the right side as τ and Δ are converging to 0. If formula 17.2 is true, then the variance of the sum in formula 17.1 can be written

$$\Delta^2 m = \frac{\Delta^2}{\tau} \cdot \tau m \to Kt.$$

The Central Limit Theorem (see Chapter 12) shows that the sum on the right-hand side of formula 17.1 when divided by the standard deviation $\Delta\sqrt{m}$ has a distribution which tends to the standard normal distribution. Since $\Delta^2 m$ converges to Kt, this will be equivalent to the distribution of the right-hand side tending to the normal distribution with mean 0 and variance Kt. The left-hand side of formula 17.1 tends to the random variable X_t. The X_t process will turn out to be Brownian motion.

To see this, we want to check that the X_t process just constructed satisfies the assumptions (a), (b), and (c) given above for Brownian motion. We have just seen that the distribution of $X_t = X_t - 0 = X_t - X_0$ is what it should be according to the definition in (b) above: it is normal with expectation 0 and variance Kt. But instead of starting from 0 at initial time 0, we could have started from 0 at any time s. Then we could have represented the difference $X_t - X_s$ as a sum of U's just as in formula 17.1 and applied exactly the same argument as before, where $t - s$ now plays the role of t above. This leads to the general statement of (b) in the definition above. The statement (a) of the definition follows from the interpretation of the random variable X_t as the sum of independent U variables. We can argue like this: Suppose three times r, s, and t are given with $r < s < t$. The

random variable $X_s - X_r$ can be written as a sum of U's as in formula
17.1, and the random variable $X_t - X_s$ can be similarly represented as
a sum of U's. But the U's in the first representation are independent of
the U's in the second representation because the time intervals from r to
s and from s to t are non-overlapping. Therefore, $X_s - X_r$ and $X_t - X_s$
are independent. The same argument works for any finite number of times
$t_1, \cdots, t_n$, and this shows (a) in the definition above is satisfied. Since we
started with $P(X_0 = 0) = 1$, the process X_t is a Brownian motion as
claimed.

The above argument shows how the Brownian motion process can be
thought of as a limit of a bunch of random walks where the unit displacement and unit time both shrink to 0. To get Brownian motion though, the
shrinking couldn't have been in any arbitrary manner. We had to impose
the conditions of formula 17.2 in order to get the variance of the sum in
formula 17.1 tending to the finite, positive number Kt. Without formula
17.2 there is no guarantee that $\Delta^2 m$ has a finite limit or any limit at all.
Then we wouldn't have been able to use the Central Limit Theorem.

From the relation $\Delta^2/\tau \to K$, we can see why the Brownian paths don't
have tangents, or to say it another way, particles moving on these paths
don't have finite velocities. Both Δ and τ are tending to 0, but Δ^2 is about
K times τ. This implies that for very small Δ and τ, Δ is much larger than
τ, so Δ/τ is getting arbitrarily large, tending to infinity. But Δ represents
the distance the Brownian particle moves in time τ, so the ratio Δ/τ is
distance divided by time, and this is the velocity of the particle. So the
limiting value of the velocity does not exist as a finite number, and the
Brownian paths have no tangents.

Since Brownian motion is the limit of random walks, and these random
walks are Markov chains, we might expect that the Markov property carries
over to Brownian motion. Recall from Chapter 16 that the essential feature
of the Markov property is independence of the present (and future) from
the "distant" past; that is, the conditional probability of a present or future
event on the past just depends on the situation at the last given time, the
present. For Brownian motion, this would say something like

$$P(X_{t_n} \text{ lies in } I/X_{t_1} = u_1, X_{t_2} = u_2, \cdots, X_{t_{n-1}} = u_{n-1})$$

only depends upon u_{n-1}, where I is any fixed interval and $t_1 < t_2 < \cdots < t_n$. To see the truth of this, note that this probability is the same as

$$P(X_{t_n} - X_{t_{n-1}} \text{ lies in } I - u_{n-1}/X_{t_1} = u_1, X_{t_2} = u_2, \cdots, X_{t_{n-1}} = u_{n-1}),$$
$$(17.3)$$

where $I - u_{n-1}$ is the interval obtained by taking the set of all numbers of
the form $x - u_{n-1}$ for all x in I. But by property (a) of the definition of
Brownian motion, $X_{t_n} - X_{t_{n-1}}$ is independent of $X_{t_1} - X_0 = X_{t_1}$ and $X_{t_2} - X_{t_1}$, and so is independent of the sum $X_{t_1} + X_{t_2} - X_{t_1} = X_{t_2}$. Therefore,
$X_{t_n} - X_{t_{n-1}}$ is independent of both X_{t_1} and X_{t_2}. By continuing in this way

we can show that $X_{t_n} - X_{t_{n-1}}$ is independent of $X_{t_1}, X_{t_2}, \cdots, X_{t_{n-1}}$. What this means is that the probability of formula 17.3 is equal to

$$P(X_{t_n} - X_{t_{n-1}} \text{ lies in } I - u_{n-1}),$$

which just depends upon u_{n-1}, the position of the particle at the last given time. This proves the Markov property for Brownian motion.

The Poisson process is also a Markov process (if we set $X_0 = 0$ with probability 1); the argument, as in the above proof for Brownian motion, depends on the fact that the Poisson process also has independent increments with the distribution of the increment only depending upon the amount of the time difference. Independent increment processes are continuous time analogs of random walks and inherit a lot of the nice properties of these walks, like the Markov property.

17.6   Exercises for Chapter 17

The processes X in the exercises below are standard Brownian motion processes.

1. Find the probability of the event:

$$X_{10} > 0, X_{21} < X_{10}, X_{25} > X_{21}.$$

2. Assume $K = 1$ in condition (b) for the Brownian motion. Express the following probability in terms of normally distributed random variable(s) (give the mean and variance):

$$P(X_{.75} > X_{.40} + 1, X_{.25} < -2).$$

3. Given an interval I, can there be a substantial number of Brownian paths remaining constant throughout the interval? Try to calculate the probability of a Brownian path remaining constant on I.

4. Assuming $K = 1$ (see exercise 2), express the probability

$$P(0 < X_8 < 1/X_4 = 2, \ -10 < X_2 < 10)$$

in terms of normally distributed random variable(s).

5. Let a be any fixed level. Show that for all times $t = T$ sufficiently large, $P(X_T < a) \approx .5$. (Hint: write the event in terms of a standard normal variable.)

6. Let $a > 0$ be any fixed level. We are interested in the event: the Brownian particle $X(t)$ stays between 0 and a forever, i.e., $0 \leq X_t < a$

for all $t \geq 0$. Using the preceding exercise, give a heuristic (that is, intuitive) argument showing the probability of this event is 0. (Hint: for all times T,

$$P(0 \leq X_t < a \text{ for all } t \leq T) \leq P(0 \leq X_T < a),$$

and take T large as in exercise 5.)

Answers to Exercises

CHAPTER 1

1. (a) Can be described as the set of all triples of the form $(D, C1, C2)$ where D stands for the number rolled on the die, and $C1$ and $C2$ stand for H or T (for heads or tails). There are 24 possible outcomes. (b) 1/8, 1/4, 1/2.

2. Assume the car is behind the door labelled 1 and the goats behind the doors labelled 2, 3, and 4. If you switch, you can only win if your initial choice is one of the doors 2, 3, or 4. Each of these choices has probability 1/4. Assume we choose door 2. Notice that whatever the host does, you wind up with two choices, one of which will win the car, and the other will win the goat. Thus, half of the probability 1/4, namely, 1/8, should correspond to the event of winning the car when you choose door 2. Since the same argument works for each of the other doors, the answer is 3/8. If you don't switch, the only way you can win the car is if your initial choice is door 1, so the answer is 1/4.

3. The wording implies that we are to consider all possible times, so there are a continuum of instants in the interval between 6 and 7 AM. The sample space is therefore continuous.

4. There are 13 possible times the alarm can go off. The sample space can therefore be represented by the 13 outcomes corresponding to the times 6:00, 6:05, 6:10, etc., up to 7:00. Because the sample space has a finite number of outcomes, it is discrete.

5. The alarm will disturb my dream if it rings at one of the five times 6:20, 6:25, 6:30, 6:35, or 6:40. The probability is 5/13.

CHAPTER 2

1. Since everyone in the elevator was born in January, there are a total of 31 days to consider. Max knows his birthday, so we look at the seven other people. The first can have a birthday chosen in 30 ways (it must be different from Max's birthday), the second in 29 ways, etc., giving the probability that all birthdays are different by

$$v = \frac{(30)(29)\cdots(24)}{(31)^7}.$$

The probability that at least two have the same birthday is $1 - v$.

2. 720 ways.

3. If we put a man in the first seat there are three choices. Then the next seat must have a woman, and there are three choices. The next seat must have a man and there are two choices, and the next a woman and there are two choices. The final two seats must have the remaining man and the remaining woman, so there is only one choice. The total number of ways if we start with a man in the first seat is therefore 36. If we start with a woman in the first seat we get another 36, for a total of 72 ways.

4. This is useful in studying sequences of heads and tails (see the discussion of the binomial distribution in Chapter 7). Assume the H's are labelled $H1, H2, H3, H4$, and the T's $T1, T2, T3$. The total number of ordered arrangements of these seven symbols is $7 \cdot 6 \cdots = 5040$. But if you want distinguishable patterns, observe that since the H's and T's cannot be distinguished among themselves, any distinguishable pattern is equivalent to 24 ordered arrangements by shuffling the H's in all possible ways, and six ordered arrangements by shuffling the T's. Therefore, we must divide 5040 by $24 \cdot 6$ to get 35 distinguishable patterns.

5. Since there are fewer numbers to choose from, there are fewer outcomes, so the chances that the number you chose is selected is greater. The number of outcomes is

$$\frac{40 \cdot 39 \cdot 38 \cdot 37 \cdot 36 \cdot 35}{720} = 3{,}838{,}380,$$

and the probability is therefore 1/3,838,380.

CHAPTER 3

1. (a) The first die is odd and the second die is even, 1/4.
(b) The first die is odd and the second is even, or else the first die is even and the second odd, 1/2.
(c) The first die is even, 1/2.
(d) The first die is even and the second odd, 1/4.

2. 1/18, 1/9, 1/8.

3. 11/16, 11/15, 1/11.

4. 3/11, 5/22, 1/2, 1/2.

5. The left-hand side of the relation can be written

$$P(A) \, \frac{P(A \cap B)}{P(A)} \, \frac{P(A \cap B \cap C)}{P(A \cap B)}.$$

Cancellation gives the right-hand side of the relation.

CHAPTER 4

1. The initial information gives you a uniform distribution for the nine possibilities of the composition of the urn (we are using a sample space with ordered pairs). Without further information, the probability of the urn containing a green ball and a red ball is 2/9 (first ball red, second green, and first ball green, second red). But once we choose a ball and note that it is green, the urn must contain a green ball. The other ball is equally likely to be either of the three colors, and the sample space can be represented by the unordered pairs $\{G, G\}$, $\{G, R\}$, and $\{G, B\}$, each with probability 1/3. The probability that the second ball chosen is red is the probability that the composition of the urn is $\{G, R\}$, which is 1/3, multiplied by the conditional probability of choosing the red ball, given this urn composition, and this is 1/2. So the answer for red is 1/6, and the symmetry of the situation (or a direct computation) gives the same answer for black. Using complementary events, we see the probability of the second ball being green must be $2/3 = 1 - 1/3$. This can be seen directly: green can be chosen under each of the three possible compositions. In the two cases where green is paired with another color, each of these cases contributes 1/6 to the total probability. In the case where there are two green balls, you choose a green ball with probability 1, so in this case, 1/3, the probability of $\{G, G\}$ contributes to the total probability, giving $2/3 = 1/6 + 1/6 + 1/3$.

2. $P(A \cap B) = 1/6$, and $P(B) = 1/2$, so the answer to (a) is $1/3$. For (b) we have

$$P(B/A) = \frac{P(A/B)\ P(B)}{P(A/B)\ P(B) + P(A/B^c)\ P(B^c)},$$

and this gives $1/6$ on top and $11/36$ on the bottom, giving the value $6/11$.

3. Use of Bayes's formula gives $(.10)(.8) = .08$ on the top, and $.08 + (.60)(.2) = .20$ on the bottom, to give a belief probability of $.4$.

4. The top of Bayes's formula has $(.40)(.5) = .20$, and the bottom has $.20 + (.10)(.5) = .25$, so the probability is $.8$.

5. In Bayes's formula (problem 2) let $B =$ winning the car, and $A =$ switching doors. The formula gives $2/3 \cdot 1/2$ on top (probability of switching is equal to $1/2$), and $2/3 \cdot 1/2 + 1/3 \cdot 1/2 = 1/2$ on the bottom. So the answer is $2/3$.

CHAPTER 5

1. The sample space can be described by the four pairs

$$(H,H), (H,T), (T,H), (T,T),$$

where the first and second entries describe whether heads or tails fall on the first (fair) coin and the second (biased) coin, respectively. The probabilities for each of the outcomes are $1/6$, $1/3$, $1/6$, $1/3$, respectively. The probability of at least one head is $2/3$, and the probability of at least one tail is $5/6$.

2. (a) $(.999)^{100}$.
   (b) 0. (c) $1 - (.999)^{1,000,000}$.

3. Does the model of repeated independent trials apply to this real-life problem? If you think so, then ask about the expected waiting time until you win. In Section 7.4 we will see that the expected waiting time until the first success in repeated Bernoulli trials is the reciprocal of p, the probability of success. For the lottery discussed in Chapter 2, this would require about 25,827,165 plays. If you play once a day every day of the year, this gives you an expected wait of about 70,759 years!

4. If you don't switch, the probability of winning a goat is $2/3$, and if you switch the probability is $1/3$, so by independence the probability of winning two goats by the strategy indicated is $2/9$. The probability of winning two cars by this strategy is also $2/9$.

5. (a) The colors of the lights as Ringo approaches can be described by a triple of the form $(C1, C2, C3)$, where $C1, C2$, and $C3$ are the colors of the first, second, and third lights, respectively. In general, we would have a sample space of eight outcomes, where we could put each of the two possible colors in for $C1, C2$, and $C3$. But in this case we only have four outcomes with positive probability:

$$(G, G, G), (G, R, R), (R, G, R), (R, R, G)$$

each with probability 1/4. (b) 1/2, 1/2, 1/2; 1/4, 1/4, 1/4; 0. If the three events were independent, then the probability of $F \cap S \cap T$ would be 1/8, not 0.

CHAPTER 6

1. Using .51 as the approximate probability of losing at a single game of craps, and assuming independence of the games at each day, gives the probability of winning at least one game equal to $1 - (.51)^4$.

2. There are three ways for my opponent to win if the game were continued, given the present score: she could win the next point; I could win the next point, and she could win the succeeding one; I could win the next two points and she could win the succeeding one. The probability of my opponent winning is therefore $1/3 + 2/3 \cdot 1/3 + 2/3 \cdot 2/3 \cdot 1/3 = 19/27$. According to Pascal's principle, she should receive $(19/27) \cdot \$100 \approx \$70.37$ and I should receive \$29.63.

3. 35:1; 1:7.

4. The game will end when you select a black ball if and only if you selected a black ball initially, and the probability of this event is 1/5. The second event means that the black ball was initially selected and the following two selections resulted in red balls. The probability is, by independence, $1/5 \cdot (4/5)^2 = 16/125 = .128$.

5. We consider sequences of plays that start immediately after Anna's winning play. The event whose probability we seek occurs if there are i consecutive plays resulting in red balls different from the red 7, followed by a play in which the red 7 appears, where $i = 0, 1, 2 \cdots$. We have to add up the probabilities of each of these possible events as i varies, and this gives the infinite series

$$1/38 + (17/38)(1/38) + (17/38)^2(1/38) + \cdots = 1/21.$$

6. No. The sample space can be represented by six ordered triples (a, b, c), where a is the number of the card, b is the color observed, and c is

the color not seen. Using a uniform distribution, the probability of observing a red color is .5, and the conditional probability of the red-red card given this observation is 2/3. The odds against the red-black card are 2 to 1.

CHAPTER 7

1. Use the binomial distribution model with success identified with rolling 7 and failure rolling anything else. Then the probability of success is 1/6, and there are 100 trials, so

$$P(X = 5) = C_{100,5} \ (1/6)^5 \ (5/6)^{95}$$

and $P(X < 98) = 1 - (P(X = 98) + P(X = 99) + P(X = 100)) =$
$1 - 4950 \ (1/6)^{98} \ (5/6)^2 - 100 \ (1/6)^{99} \ (5/6) - (1/6)^{100}.$

2. The probability that the number you bet on appears at least once is $1 - (5/6)^3 = Q$. Then the conditional probabilities that the number bet on appears 1, 2, and 3 times are

$$(3 \cdot 1/6 \ (5/6)^2)/Q, \ (3 \cdot (1/6)^2 \ 5/6)/Q, \ \text{and} \ (1/6)^3/Q,$$

respectively, giving approximate values of .824, .164, and .012. (Notice these numbers add up to 1, as they must.) The expectation is calculated by multiplying each of the conditional probabilities by 1, 2, and 3, respectively, and then adding up. This gives an approximate expectation of $1.19.

3. The expected payoff to you of the game is

$$2 \cdot (2^{-1}) + 2^2 \cdot (2^{-2}) + \cdots + 2^N \cdot (2^{-N}) = N.$$

The expected payoff should be equal to the entrance fee to play the game in a fair game, so the entrance fee should be N.

4. Use the Bernoulli trial set-up with success interpreted as rolling 7, failure for anything else. The probability of success is 1/6, so the expected number of trials before success appears is the reciprocal of the success probability, 6. In the game, you win if there is at least one 7 in three rolls, and the probability is $1 - (5/6)^3$. Your expected loss is $-3 \cdot (5/6)^3$. If P is your payoff if you win, to make the game fair $P \ (1 - (5/6)^3) - 3 \cdot (5/6)^3 = 0$. Solving for P, we get an approximate value of $4.14.

5. We'll solve this problem by a direct computational approach, but in exercise 4 of Chapter 8 a slicker approach will be given which requires no computation. We consider a sample space whose outcomes are

all possible choices of two cards from the deck. The sample space can either consist of the ordered choices or the unordered ones; you will of course get the same answer (if you do the problem correctly) whichever sample space you choose. Since order is not important in this problem, just the hand one obtains, let's take the unordered choices. Then the sample space contains $C_{52,2}$ outcomes. The number of ways of choosing exactly one black card is $C_{26,1} \cdot C_{26,1}$, where the first term is the number of ways of choosing one black card from 26 possibilities, and the second term is the number of ways of choosing one red card from 26 possibilities. The number of ways of choosing two black cards is $C_{26,2}$. The expected number of black cards is then

$$1 \cdot \frac{(C_{26,1})^2}{C_{52,2}} + 2 \cdot \frac{C_{26,2}}{C_{52,2}},$$

which, if you substitute numbers and work it out, comes out to be 1. To find the expected number of hearts, observe that exactly one heart is obtained by choosing the heart in $C_{13,1}$ ways, and the non-heart in $C_{39,1}$ ways. Exactly two hearts can be chosen in $C_{13,2}$ ways. This gives the expected number of hearts to be

$$1 \cdot \frac{C_{13,1} \, C_{39,1}}{C_{52,2}} + 2 \cdot \frac{C_{13,2}}{C_{52,2}},$$

which works out to be 1/2.

6. When the balls are replaced, the choices are independent (we assume the balls are mixed after each replacement). We can use a binomial distribution, where we interpret a red ball as success for each of three trials, and with success probability .6. The expected number of red balls is then

$$1 \cdot (3)(.6)(.4)^2 + 2 \cdot (3)(.6)^2(.4) + 3 \cdot (.6)^3,$$

or 1.8 [in Section 8.2 we will see that a binomial variable always has expected value = number of trials multiplied by the success probability—this translates here into $3 \cdot (.6) = 1.8$]. Now suppose we choose without replacement. Then the probabilities of 1, 2, and 3 red balls are given by

$$\frac{C_{6,1} \cdot C_{4,2}}{C_{10,3}}, \; \frac{C_{6,2} \cdot C_{4,1}}{C_{10,3}}, \; \frac{C_{6,3}}{C_{10,3}}.$$

These are examples of probabilities from a *hypergeometric* distribution. You can calculate the expected number of red balls obtained by choosing without replacement by multiplying the probabilities above by 1, 2, and 3, respectively, and adding up. We get 1.8, the same as the expected number choosing *with replacement.* Is this just a coincidence? The answer is no! The following interesting fact can be proved: if you have r red balls and b black balls in an urn, and you make s drawings from the urn, where s must, of course, be at most equal to the total number of balls $r + b$, then the expected number of red balls selected in the s drawings is the same whether you choose with replacement or without replacement. The proof consists in finding the expected value of a hypergeometric distribution and recognizing it as the same number obtained in the case of choosing with replacement.

CHAPTER 8

1. The neatest way to do this is to let X be the first integer picked and Y be the number rolled by the die, and then use the formula: $E(X + Y) = EX + EY$. Here $EX = 2/3$, and EY is the sum of the integers 1 through 6 divided by 6, which is 3.5, and so $E(X + Y) \approx .67 + 3.50 = 4.17$. An alternative way is to consider the sum $X + Y$ as a random variable U, say, whose distribution we will derive. If 0 is the first integer picked, then U takes on the values 1 through 6, each with probability $(1/3)(1/6) = 1/18$. If 1 is the first number picked, then U takes on the values 2 through 7, each with probability $(2/3)(1/6) = 2/18$. The distribution of U can be given by the list: 2, 3, 4, 5, and 6 occur with probability $3/18 = 1/6$; 1 occurs with probability 1/18, and 7 with probability 2/18. EU is computed by taking each value, multiplying by the probability of this value, and adding up. We obtain $20/6 + 1/18 + 14/18 \approx 4.17$.

2. Let X, Y, and Z be 1 or 0 depending on whether the first, second, or third comes up heads or tails, respectively. Then $EX = 1/2, EY = 2/3$, and $EZ = 3/4$. The expected total number of heads is therefore $1/2 + 2/3 + 3/4 \approx 1.92$.

3. Think of rolling 7 as a success, anything else a failure. We can use a binomial distribution model with probability of success $= 1/6$. Let T_1 be the time until the first success, T_2 the time starting right after the first success until the second success, etc. The expected time until the tenth success is given by the expectation of

$$T_1 + T_2 + \cdots + T_{10}.$$

The variables T all have the same distribution as waiting time variables until the first success; this is true because after any success the process "starts afresh," that is, it is independent of the past. Since the expectation of each T is the reciprocal of the success probability, each T has expectation 6, so the expected time until the tenth success is 60 rolls of the dice.

4. Use the hint: if you choose two cards, then the number of black cards plus the number of red cards you hold $= 2$, that is $B + R = 2$. By symmetry, B and R have the same distribution: there are 26 black cards and 26 red ones. Therefore, B and R must have the same expectations. Then using the relation $EB + ER = E(B + R) = E2$, we get

$$EB + ER = 2, \quad \text{or } 2EB = 2, \text{ so } EB = ER = 1.$$

A similar argument works with the expected number of hearts. Let H, D, C, and S stand, respectively, for the number of hearts, diamonds, clubs, and spades chosen when you select two cards at random. We must have $H + D + C + S = 2$. As before, the symmetry (each suit has the same number of cards) implies each variable has the same distribution and therefore the same expectation. This leads to $EH + ED + EC + ES = 2$, or $4EH = 2$, from which follows $EH = ED = EC = ES = 1/2$.

5. (a) Use the relative frequency of appearances of black in a large number of plays at roulette, (b) relative frequency of rolling snake eyes in a large number of throws of a pair of dice, (c) relative frequency of winning at least \$1 in a large number of plays at chuck-a-luck, (d) relative frequency of winning in a large number of plays at the car-goat game if your strategy is always to switch.

6. By the strong Law of Large Numbers, for all n sufficiently large we must have the relation

$$P\left(\frac{S_n}{n} - .01 > -.001\right) > .99,$$

where S_n is our total winnings after n games. A little algebra shows the event in question is equivalent to $S_n > (.01 - .001)n = .009n$; that is, the total winnings will eventually exceed \$.009n with probability exceeding .99. So (b) is always true. On the other hand, it also follows in a similar manner that the relation $S_n < (.01 + .001)n$ will be true for all n sufficiently large with probability exceeding .99. This means that the total winnings will be less than $.02n < n$ dollars with probability exceeding .99 (and with probability tending to 1 as n increases beyond all bounds). This means that (a) will always be false for all sufficiently large n.

CHAPTER 9

1. (a) $e^{-5} 5^6/6!$. (b) $1 - (1 - e^{-5})^{10}$. (c) $(1 - e^{-5})^{10}$.
   (d) $10 \cdot (1 - 37/2 \, e^{-5})(37/2 \, e^{-5})^9$.

2. $.5\, e^{-1}$.

3. $e^{-30/4}$.

4. $e^{-20}\, 20^{30}/30!$.

5. The probability of the insect laying r eggs of which exactly k survive is

$$P(r \text{ eggs laid}) \cdot P(k \text{ of the } r \text{ eggs survive}/r \text{ eggs laid}),$$

where the second probability is based on the binomial model, so we get

$$e^{-5}\, 5^r/r! \cdot C_{r,k} p^k q^{r-k}.$$

The answer is obtained by adding up all of these terms for all the possible values of r, namely, $r = k, k+1, \cdots$.

(b) We need to calculate probabilities conditioned on our knowledge of the event A = at most 3 eggs were laid. We have

$$P(A) = e^{-5} + 5e^{-5} + (25/2)e^{-5} + (125/6)e^{-5} = Q.$$

The probability we seek is then

$$\frac{1}{Q}\left(5e^{-5}p + 25e^{-5}pq + (125/2)e^{-5}pq^2\right).$$

CHAPTER 10

1. We estimate your probability q of losing \$1 at .51, and p, the probability of winning \$1 at .49 (see Chapter 6). Formula 10.6 gives us the approximate value

$$\frac{(1.04)^3 - (1.04)^6}{1 - (1.04)^6} \approx .54$$

of your ruin. Since the game is unfavorable for you, the best strategy you can use, i.e., the one to reduce your ruin probability as much as possible, is to play at \$3 stakes, the largest possible. A calculation shows this reduces the ruin probability to almost .5.

2. Here $q = 20/38, p = 18/38$, and the formula gives

$$\frac{(10/9)^3 - (10/9)^6}{1 - (10/9)^6} \approx .58.$$

3. This is a fair game, so we use formula 10.8, and at \$1 stakes we obtain a ruin probability of $1 - i/2i = 1/2$. If the stakes are changed allowing the boldest play, that is betting $\$i$ at each play, the formula becomes $1 - 1/2 = 1/2$, so there is no change in ruin probability. If p is changed to .499, the game is unfavorable and the best strategy is now bold play: betting at $\$i$ stakes. If p is changed to .501, then the game is favorable, and the best strategy is timid play: bet at the lowest possible stakes allowable.

4. Ginger can be ruined after the first play with probability 4/5. She can be ruined at the third play by winning, losing, and losing again: this probability is $(4/5)(4/25)$. She can be ruined at the fifth play by oscillating between 5 and 10 and then losing, with probability $(4/5)(4/25)^2$. The general pattern is that we can oscillate between 5 and 10 any fixed number of times before being ruined. Each of these oscillations followed by ruin defines a disjoint sequence of events, so that the total probability of ruin is obtained by adding the following infinite series:

$$4/5 + (4/5)(4/25) + (4/5)(4/25)^2 + (4/5)(4/25)^3 + \cdots.$$

This is a geometric series with ratio 4/25 and initial term 4/5. The sum of the series is therefore 20/21. Use of formula 10.6 gives

$$(4 - 4^3)/(1 - 4^3) = 60/63 = 20/21,$$

where the exponents 1 and 3 used in the formula are due to the \$5 stakes.

5. If $s = 1$, we have the classical game. If $s > 1$, then we can win more than we can lose at a single play and intuitively it should become easier to win the game. This would increase the gambler's win probability and decrease his ruin probability. To make this intuitive argument a little more rigorous, note that any sequence of plays leading to the gambler winning in the classical game corresponds to another sequence of plays leading to the gambler winning in the revised game. The second sequence is an initial piece of the first sequence in general because every time we win a play we move more units to the right so the game ends faster. Since there are fewer steps to win, there are fewer probabilities to multiply together, and the sequence of plays in the revised game would have a higher probability. By adding all the probabilities corresponding to these sequences we would get the (total) probability of winning in the classical game, and by the correspondence this would not be greater than the probability of winning in the revised game. It follows that the probability of ruin in the revised game is not greater than the probability of ruin in the classical game.

CHAPTER 11

1. For Felix and Alice to be on the same bus they would both have to arrive during one of the four 15-minute intervals preceding the last four bus departures. The probability that each will arrive in a given 15-minute interval is $1/4$; that both will arrive is $1/16$. The answer is therefore $4/16 = 1/4$.

2. Let's say the stick is of unit length and goes from left endpoint A to right endpoint B. By elementary algebra we see that the stick is broken in two pieces such that one piece is more than twice as long as the other if the point of breakage is anywhere up to $1/3$ of the way from A to B or $1/3$ of the way from B to A. The probability of falling into this region is $2/3$.

3. Given any point Q in a circle, it uniquely defines a chord for which Q is the midpoint by drawing the line OQ from the center O of the circle to Q, and then drawing the chord to be perpendicular to OQ through Q. Let us think of choosing the point Q uniformly in the circle (whose radius we can assume is 1) and constructing the chord. From elementary geometry we see that the chord will exceed the side of the inscribed triangle if Q lies on the radius a distance less than $1/2$ unit from O. So if Q is selected from a circle of radius $1/2$ with center O, it gives rise to a chord at least as large as a side, but outside this circle the chord does not have this property. The probability is therefore the ratio of the areas of the two circles: $1/4$.

4. It's best to draw a picture before doing this problem. In analogy with the first solution for the triangle, we select a point Q at random on a radius and then draw the chord through Q perpendicular to the radius. The length of the chord exceeds a side of the square if and only if Q lies inside the square. Assuming the radius of the circle is 1, Q is inside the square if it lies within $\sqrt{2}/2 \approx .71$ of the center, and this is the probability sought. In analogy with our second solution for the triangle, draw a tangent to the circle at a vertex V of the square, and consider all chords of the circle having V as an endpoint. Any such chord makes an angle between 0 and 180 degrees with the tangent. The chord is larger than the side of the square if the chord falls into the square, and this occurs if the chord falls within a 90-degree arc. The probability is $1/2$.

5. Yes. The assertion of normality is a statement about the behavior of the tail part of the infinite decimal, that is, about all digits excluding at most a finite set of them. Multiplying a normal number by a positive power of 10 just shifts the decimal point a finite number of places to the right. The decimal part of this number just omits a finite number of digits from w and must be normal. On the other hand, if

w is not normal, omitting a finite number of digits by multiplying by a power of 10 produces a decimal that is also not normal.

6. (a) 0, (b) .16, (c) .19.

CHAPTER 12

1. The event $130 < 2X < 136$ is equivalent to the event $65 < X < 68$. Subtracting the mean from X and dividing by the standard deviation gives the equivalent event $-1 < (X - 67)/2 < .5$, or equivalently, $-1 < Z < .5$ for a standard normal Z.

2. The standardization of S_n is

$$\frac{S_n - n/2}{(1/2)\sqrt{n}}.$$

The above standardization will be approximately standard normal for large n by the Central Limit Theorem. So if we put

$$\frac{S_n - n/2}{(1/2)\sqrt{n}} \approx Z,$$

we can write (algebra!)

$$R_n = \frac{S_n}{\sqrt{n}} \approx \frac{Z}{2} + \frac{\sqrt{n}}{2}.$$

It follows that the probability of the event $R_n < x$ will have probability very close to the event $Z < 2x - \sqrt{n}$ for n sufficiently large. Since the right-hand side of this inequality becomes smaller than any fixed negative number as n increases, the probability of this event shrinks to 0 as n gets large.

3. S_n is the sum of independent variables X_i, which are 1 or -1 on the ith play of the game with probabilities p and q, respectively. Then X_i has expectation $p - q$ and variance $1 - (p-q)^2$ and S_n has expectation $n(p - q)$ and variance $n(1 - (p - q)^2)$. It follows that

$$\frac{S_n - n(p - q)}{\sqrt{n}\sqrt{1 - (p - q)^2}}$$

has approximately the standard normal distribution for n sufficiently large.

4. We can certainly find $x > 0$ so large that the standard normal Z satisfies $P(Z < x) > 1 - \varepsilon$. If we put in for Z the approximation given above and do a little algebra, we get

$$P(S_n < n(p - q) + \sqrt{n} \cdot K) > 1 - \varepsilon,$$

where K is the constant given by $\sqrt{1 - (p - q)^2} \cdot x$.

5. Since the game is unfavorable for the gambler, $p - q < 0$. According to the hint, this means as n increases S_n becomes more and more negative with probability at least $1 - \varepsilon$.

6. The random variable U is also standard normal. We can see this from the following equalities: $P(U < x) = P(-Z < x) = P(Z > -x) = P(Z < x)$. First think of x as positive; then the last equality is clear from symmetry. If you then think of x as negative, again symmetry shows the last equality. Since U and Z have the same cumulative distribution function (so their probability distributions agree on half infinite intervals of the form $\{w: w < x\}$, it will follow that U and Z have the same distribution (their distributions agree on all intervals).

CHAPTER 13

1. One way to assign patients to a group randomly is to choose a random number as each patient enters. If the number is even, the patient goes into the treatment group with the new drug; if the number is odd, he goes into the group using the old drug. This method has the disadvantage that the groups may have very different numbers of patients. To assure that each group gets 20 patients, each of 40 prospective participants is assigned a number 00 to 39. We will choose random numbers from a table, and decide in advance that the first 20 numbers chosen from the set 00 to 39 will go into the treatment group with the new drug, say, and the remaining patients will go into the other group. In the table of random numbers, we choose pairs, and only select those in the range 00 to 39, ignoring all others. Stop when 20 numbers have been selected; the patients assigned these numbers go into the group treated by the new drug.

2. Select a random digit; if it is even, select 0, and if odd, select 1. Another way: select a random digit. If it is 0 through 4, select 0, otherwise select 1. Each of these methods chooses each of the digits 0 and 1 with probability 1/2. To choose 0, 1, or 2 with probability 1/3, one way would be to select a random digit 0 through 8 (ignore 9). If it is 0, 1, or 2, select 0; if it is 3, 4, or 5, select 1. Otherwise select 2.

3. A random number is one that has been produced by a chance mechanism, so it has used chance as an intrinsic part of its definition. A pseudo-random number is deterministic; it does not use random mechanisms for its generation. It has, however, been constructed such that its statistical properties are very similar to those of random numbers.

4. About 10 times in 1000.

5. The total area of the square is 9 square units. The Monte Carlo principle assumes that the relative frequency of points in R is roughly proportional to the relative areas of R and the square. This would give us the value $(.35)(9) = 3.15$ square units as the estimate for the area of R.

CHAPTER 14

(We only sketch the idea for these problems instead of giving full algorithms. By studying the examples of Chapter 14, you should be able to convert our sketchy descriptions into proper algorithms.)

1. (a) Enter the number you bet on. Generate three independent rolls of a die. Count how many times the number bet on came up on the dice. Pay the gambler this amount. (b) Repeat the game many times. Count the number of times the gambler wins $2 in the game. Divide this number by n, the total number of repetitions of the game, to get the relative frequency of winning $2. This is an estimate of the probability.

2. Identify the digits 1, 2, and 3 with car, goat, goat, respectively. Choose one of the digits, X, at random (player's initial choice). If $X = 1$, then the player wins if she doesn't switch and loses if she switches. If $X = 2$ or $X = 3$, then the player loses if she doesn't switch and wins if she switches. To estimate the probability of winning if she switches, use a counter to count for each play of a large number, n, of games, the number of times she wins when switching. Divide the counter by n to get the relative frequency of wins when switching; this estimates the desired probability.

3. Only the instructions 9 and 11 need to be changed. In 9, if we replace "$Y > -1$ and $Y < 1$" with "$Y > .5$" and adjust 11 in an obvious way, we estimate $P(Y > .5)$. If we replace with the event $-.3 < Y < .3$, and adjust 11, we estimate the probability of this event. In each case, we can compare with a table for the standard normal distribution to see how close the distribution of Y is to the limiting one.

4. To simulate roulette, first identify 38 random numbers, let's say the integers 1 to 38, with the outcomes of a roulette wheel. For example, 1 to 18 could represent the black numbers, 19 to 36 the red ones, and 37 and 38 the 0 and 00 values. Now choose one of the 38 numbers randomly to determine the roulette outcome. To estimate the probability of winning by playing black, repeat the game a large number, n, of times and count the number of times one of the numbers from 1 to 18 turns up (we are using the representation above; your method may

be different). Divide the value of the count by n to get the relative frequency of black winning. This is an estimate of the probability.

5. We want to choose a point uniformly distributed on the square. To do this, choose two values X and Y independently and uniformly on the interval from 0 to 1. The point (X, Y) then represents a point randomly chosen on the square. The point lies between the two curves if and only if the inequalities $X^3 < Y$ and $Y < X^2$ are both satisfied. Choose a large number, n, of random points in this way, and define a counter to count the frequency of points satisfying both inequalities. This frequency divided by n is the relative frequency of points chosen at random from the square that fall between the two curves. This ratio should be close to the ratio of the desired area to 1, the total area of the square. So the estimate of the desired area is obtained by obtaining the relative frequency of points falling in this area.

6. We choose 100 numbers to represent the 100 coins; the easiest way is to let 01 to 99 represent the coins 1 to 99 and let 00 represent coin 100. If X is one of the numbers chosen at random, then if $X = k$, a number 1 to k is chosen at random (if $k = 00$ we make the obvious adjustment). Let 1 represent heads, and the other $k - 1$ numbers tails. If k is even and 1 is chosen, then a counter G (initialized to 0) is increased by 1. If k is odd, then G is increased by 1 if 1 does not turn up. Repeat the play a large number, n, of times. G counts the number of times Guildenstern wins \$1. Divide the value of G by n to get the relative frequency. This is an estimate of the probability of Guildenstern's winning \$1 in a single play.

CHAPTER 15

1. Let h be the number of successes in the 100 trials. The test is based on the inequality

$$(.33)^h(.67)^{100-h} < (.67)^h(.33)^{100-h}.$$

The principle of maximum likelihood tells us to choose $p = 2/3 \approx .67$ whenever this inequality is satisfied, and $p = 1/3$ when the reverse inequality is satisfied. A calculation shows the above inequality is satisfied for $h > 50$.

2. Let n be a large number of trials, and let h be the total number of successes. Compute the value of exactly h successes under each of the competing N probabilities. Estimate p to be the probability giving a maximum value among the N competing probabilities (if the maximum value is not unique, choose among the probabilities giving this value randomly, or repeat the experiment until there is a unique maximum).

3. If H_0 is true, the random variable

$$\frac{S_n/n - 1/6}{\sqrt{(1/6) \cdot (5/6) \cdot (1/n)}}$$

is approximately standard normal. Put in the numbers in the above (900 for S_n and 6000 for n). We obtain approximately -3.5. The probability of a standard normal variable Z taking on a value that far from 0 is less than .002, a value small enough that we would in general be inclined to reject H_0, concluding that the dice are not fair.

4. We obtain the confidence interval $.46 \pm 1.96 \sqrt{(.46)(.54)/1000} \approx .46 \pm$ .03. If Groucho needs at least 50 percent of the vote to win, this result is not too encouraging. Since the true proportion of voters is included in the confidence interval 95 percent of the time such an interval is computed, and the computed interval here is $(.43, .49)$ which excludes the minimum value of .50, there is cause for concern for Groucho's supporters. If 480 rather than 460 plan to vote for Groucho, the interval becomes $(.45, .51)$, better news for Groucho because the interval now covers the critical value .50. Of course, the true value of the parameter can still be $< .50$.

5. The assumptions are the same as in Section 15.4. The ratio of tagged fish to total fish caught in the lake after the initial tagging is $250/400$. We assume that this ratio between tagged and untagged fish holds in the entire lake. Therefore, $800/x = 250/400$, where x represents the total number of fish in the lake. We get the estimate $x = 1280$ fish in the lake.

6. We give relative frequencies of numbers or sets of numbers coming up as estimates for the corresponding probabilities. (a) .19, (b) .25, (c) .12, (d) .06, (e) .64.

CHAPTER 16

1. $p = 18/38, q = 20/38$, so the probability is $(9/10)^5$.

2. We must return to 0 in 4 steps. So if a and b are the number of steps to the right and to the left, respectively, then $a = b$ and $a + b = 4$. This means there are two steps to the right and two steps to the left. There are six paths possible, so the answer to (a) is $6p^2q^2$. The answer to (b) follows from the relations $a = b$ and $a + b = n$, which must hold if the walk starts at 0 and is again at 0 at step n. This implies $2a = n$, and n must therefore be even.

3. Suppose there exists a set of paths S (with positive probability) from a that never return to 0 (that is, 0 is not an entry in the sequence defining a path). Consider the shortest path, T, from 0 to a with positive probability. Such a path exists by assumption. If we traverse T followed by a path of S, we have produced a path from 0 that never returns to 0. The set of such paths has positive probability because

$$P(X_n \text{ never returns to } 0 \, / X_0 = 0)$$
$$\geq P(T/X_0 = 0) \cdot P(S/X_n = a) > 0,$$

so the chain can escape from 0 with at least this much probability, contradicting the assumption of the chain returning to 0 with probability 1.

4. Assume a stationary distribution exists and let i be a state where $v(i)$ is a maximum. We must have the relation $v(i-1)p + v(i+1)q = v(i)$. The left-hand side of this relation is an average of two values, and an average is always less than one of the values averaged unless the values and the average are all equal. Since neither $v(i-1)$ nor $v(i+1)$ can be larger than $v(i)$, these three values must all be equal. So the neighbors of $v(i)$ also have the maximum value. By continuing this argument, we see that all states must have the maximum value, which is impossible since there are an infinite number of states, and the sum of the v values must be 1.

5. For any state $i > 0$, we have $(v(i-1)).5 = v(i)$. We also have

$$(v(0) + v(1) + v(2) + \cdots) \, .5 = v(0).$$

Moreover, the sum of all the v values is 1, so the relation above shows $v(0) = .5$; then from the first relation we get for all i, $v(i) = 2^{-i-1}$, and this is a stationary distribution for the chain.

CHAPTER 17

1. The event can be written

$$X_{10} > 0, X_{21} - X_{10} < 0, X_{25} - X_{21} > 0.$$

The three conditions defining this event are independent, and each of the variables $X_{10}, X_{21} - X_{10}$, and $X_{25} - X_{21}$ are normal with expectation 0. The probability of each of these variables exceeding 0 is therefore $1/2$, and the event has probability $1/8$.

2. The answer is

$$P(X_{.75} - X_{.40} > 1)P(X_{.25} < -2),$$

where $X_{.75} - X_{.40}$ is normal with mean 0 and variance equal to .35, and $X_{.25}$ is normal with mean 0 and variance equal to .25. The answer can be written as a product because the variables are independent.

3. Since a typical Brownian path is an erratic creature, constantly changing direction, intuition suggests the answer is "no." To see this more rigorously, let a and b be the left and right endpoints of I. Since the path is constant throughout the interval, we will have $X_a = X_b$ or, equivalently, $X_b - X_a = 0$. But $X_b - X_a$ is normal, so the probability that it takes on any single value is 0. It follows that the probability of the path remaining constant on I is 0.

4. The probability can be written

$$P(-2 < X_8 - X_4 < -1/\ X_4 = 2, -10 < X_2 < 10).$$

The variable $X_8 - X_4$ is independent of the conditioning events, so the answer is $P(-2 < X_8 - X_4 < -1)$, where $X_8 - X_4$ is normal with mean 0 and variance 4.

5. Assume $K = 1$ and standardize X_T to get a standard normal variable $X_T/\sqrt{T} = Z$. Then the event $X_T < a$ is equivalent to the event $Z < a/\sqrt{T}$. For large enough T, $a/\sqrt{T}$ is close to 0, so

$$P(Z < a/\sqrt{T}) \approx P(Z < 0) = 1/2.$$

The approximation becomes better the larger T becomes.

6. The variable X_T has mean 0, so $P(X_T < 0) = 1/2$. In the preceding exercise, we have seen that $P(X_T < a) \approx 1/2$ for large T. Since

$$P(0 \le X_T < a) = P(X_T < a) - P(X_T < 0),$$

we see that the left-hand side of this relation tends to 0 as T increases. Now use the hint.

On the following several pages you will find a short list of books and articles about probability and related areas. This list is not meant to be complete in any way; it is only a small selection of works I believe may draw the reader further into the subject. Some of the references have been cited in this book. Most of them are fairly accessible to those with modest mathematical backgrounds. Next to a number of the entries I have included brief comments as a rough guide to the reader.

Bibliography

[1] Bass, Thomas A., *The Eudaemonic Pie*, Houghton-Mifflin, Boston, 1985.

[2] Browne, Malcolm, "Coin-Tossing Computers Found to Show Subtle Bias," *New York Times*, Jan. 12, 1993.

[3] Chung, K.L., *Elementary Probability Theory with Stochastic Processes*, Springer-Verlag, New York, 1979. (A rather difficult introductory text.)

[4] Davis, Morton D., *Game Theory: A Nontechnical Introduction*, Basic Books, New York, 1970.

[5] Davis, Morton D., *Mathematically Speaking*, Harcourt Brace Jovanovich, New York, 1980.

[6] Davis, Morton D., *The Art of Decision Making*, Springer-Verlag, New York, 1986.

[7] Dowd, Maureen, "People Are Ready For Sacrifices, Poll Finds, and Expect Fairness," *New York Times*, Feb. 16, 1993.

[8] Feller, William, *An Introduction to Probability Theory and Its Applications*, Wiley, New York, 1960. (This is a classic first text, but assumes a mathematically oriented reader.)

[9] Fienberg, Stephen E., "Randomization and Social Affairs: The 1970 Draft Lottery," *Science*, Vol. 171, 1971, pp. 255-261.

[10] Fienberg, Stephen E. (editor), *The Evolving Role of Statistical Assessments as Evidence in the Courts*, Springer-Verlag, New York, 1989.

[11] Freedman, D., Pisani, R., and Purves, R., *Statistics*, Norton, New York, 1978. (A very elementary text in statistics.)

[12] Freund, John E., *Introduction to Probability*, Dover, New York, 1993. (An elementary introduction.)

[13] Gillman, Leonard, "The Car and the Goats," *American Mathematical Monthly*, Vol. 99, 1992, pp. 3-7.

[14] Goldberg, Samuel, *Probability: An Introduction*, Dover, New York, 1960.

[15] Hamming, Richard W., *The Art of Probability for Scientists and Engineers*, Addison-Wesley, Reading, Mass., 1991.

[16] Hodges, Jr., J.L., and Lehmann, E.L., *Elements of Finite Probability*, Holden-Day, San Francisco, 1965.

[17] Hodges Jr., J.L. and Lehmann, E.L., *Basic Concepts of Probability and Statistics*, Holden-Day, San Francisco, 1966.

[18] Iosifescu, Marius, *Finite Markov Processes and Their Applications*, Wiley, New York, 1980.

[19] Isaac, Richard, "Cars, Goats, and Sample Spaces: A Unified Approach," *Mathematics in College* (City University of New York), Fall-Winter 1993.

[20] Kac, Mark, "Random Walk and The Theory of Brownian Motion," *American Mathematical Monthly*, Vol. 54, 1947, pp. 369-391.

[21] Kac, Mark, *Statistical Independence in Probability, Analysis and Number Theory*, Mathematical Association of America; distributed by Wiley, 1959. (An interesting little book for the mathematically sophisticated; need calculus and concentration.)

[22] Keynes, J.M., *A Treatise on Probability*, Macmillan, London, 1921. (A classic.)

[23] Laplace, Pierre Simon, Marquis de, *A Philosophical Essay On Probabilities*, Dover, New York, 1951.

[24] von Mises, Richard, *Probability, Statistics, and Truth*, Allen and Unwin, London, 1957. (A book on the foundations of probability from the frequentist viewpoint.)

[25] Mosteller, Frederick, *Fifty Challenging Problems in Probability with Solutions*, Dover, New York, 1965. (Statements and brief solutions of famous probability problems, some of which also appear in this book.)

[26] Niven, Ivan, *Irrational Numbers*, Mathematical Association of America; distributed by Wiley, 1956. (Contains a discussion of normal numbers; for the mathematically oriented.)

[27] Rand Corporation, *A Million Random Digits with 100,000 Normal Deviates*, The Free Press, New York, 1955.

[28] Ross, Sheldon, *A First Course in Probability*, Macmillan, New York, 1988. (Good, basic modern introductory text for those with a strong mathematical background.)

[29] Savage, Leonard J., *The Foundations of Statistics*, Wiley, New York, 1954. (From a subjectivist's perspective; for the mathematically sophisticated.)

[30] Scarne, John, *Scarne's Guide to Casino Gambling*, Simon and Schuster, New York, 1978. (Rules of casino games interlaced with anecdotes and advice.)

[31] Sullivan, Joseph F., "Paternity Test at Issue in New Jersey Sex-Assault Case," *New York Times*, Nov. 28, 1990.

[32] Tierney, John, "Behind Monty Hall's Doors: Puzzle, Debate, and Answer?," *New York Times*, July 21, 1991.

[33] Thorp, Edward, *Beat the Dealer*, Random House, New York, 1962.

[34] Weatherford, R., *Philosophical Foundations of Probability Theory*, Routledge & Kegan Paul, London, 1982.

[35] Weaver, Warren, *Lady Luck*, Anchor Books-Doubleday & Co., Garden City, N.Y., 1963 (A popular classic; easier than this book. Reprinted by Dover, New York.)

Index

Undergraduate Texts in Mathematics

Anglin: Mathematics: A Concise History and Philosophy.
Readings in Mathematics.

Anglin/Lambek: The Heritage of Thales.
Readings in Mathematics.

Apostol: Introduction to Analytic Number Theory. Second edition.

Armstrong: Basic Topology.

Armstrong: Groups and Symmetry.

Axler: Linear Algebra Done Right. Second edition.

Beardon: Limits: A New Approach to Real Analysis.

Bak/Newman: Complex Analysis. Second edition.

Banchoff/Wermer: Linear Algebra Through Geometry. Second edition.

Berberian: A First Course in Real Analysis.

Bix: Conics and Cubics: A Concrete Introduction to Algebraic Curves.

Brémaud: An Introduction to Probabilistic Modeling.

Bressoud: Factorization and Primality Testing.

Bressoud: Second Year Calculus.
Readings in Mathematics.

Brickman: Mathematical Introduction to Linear Programming and Game Theory.

Browder: Mathematical Analysis: An Introduction.

Buskes/van Rooij: Topological Spaces: From Distance to Neighborhood.

Callahan: The Geometry of Spacetime: An Introduction to Special and General Relativity.

Carter/van Brunt: The Lebesgue–Stieltjes Integral: A Practical Introduction

Cederberg: A Course in Modern Geometries.

Childs: A Concrete Introduction to Higher Algebra. Second edition.

Chung: Elementary Probability Theory with Stochastic Processes. Third edition.

Cox/Little/O'Shea: Ideals, Varieties, and Algorithms. Second edition.

Croom: Basic Concepts of Algebraic Topology.

Curtis: Linear Algebra: An Introductory Approach. Fourth edition.

Devlin: The Joy of Sets: Fundamentals of Contemporary Set Theory. Second edition.

Dixmier: General Topology.

Driver: Why Math?

Ebbinghaus/Flum/Thomas: Mathematical Logic. Second edition.

Edgar: Measure, Topology, and Fractal Geometry.

Elaydi: An Introduction to Difference Equations. Second edition.

Exner: An Accompaniment to Higher Mathematics.

Exner: Inside Calculus.

Fine/Rosenberger: The Fundamental Theory of Algebra.

Fischer: Intermediate Real Analysis.

Flanigan/Kazdan: Calculus Two: Linear and Nonlinear Functions. Second edition.

Fleming: Functions of Several Variables. Second edition.

Foulds: Combinatorial Optimization for Undergraduates.

Foulds: Optimization Techniques: An Introduction.

Franklin: Methods of Mathematical Economics.

Frazier: An Introduction to Wavelets Through Linear Algebra.

Gordon: Discrete Probability.

Hairer/Wanner: Analysis by Its History.
Readings in Mathematics.

Halmos: Finite-Dimensional Vector Spaces. Second edition.

Halmos: Naive Set Theory.

Hämmerlin/Hoffmann: Numerical Mathematics.
Readings in Mathematics.

Harris/Hirst/Mossinghoff: Combinatorics and Graph Theory.

Hartshorne: Geometry: Euclid and Beyond.

Hijab: Introduction to Calculus and Classical Analysis.

Undergraduate Texts in Mathematics

Thorpe: Elementary Topics in Differential Geometry.

Toth: Glimpses of Algebra and Geometry. *Readings in Mathematics.*

Troutman: Variational Calculus and Optimal Control. Second edition.

Valenza: Linear Algebra: An Introduction to Abstract Mathematics.

Whyburn/Duda: Dynamic Topology.

Wilson: Much Ado About Calculus.

9781461269120